AF736598

Digital Deliverance

Dragging Rural America, Kicking and Screaming, into the Information Economy

Judson C. Edwards

UNIVERSITY PRESS OF AMERICA,® INC.
Lanham • Boulder • New York • Toronto • Plymouth, UK

Copyright © 2009 by
University Press of America,® Inc.
4501 Forbes Boulevard
Suite 200
Lanham, Maryland 20706
UPA Acquisitions Department (301) 459-3366

Estover Road
Plymouth PL6 7PY
United Kingdom

All rights reserved

British Library Cataloging in Publication Information Available

Library of Congress Control Number: 2009924199
ISBN: 978-0-7618-4552-2 (paperback : alk. paper)
eISBN: 978-0-7618-4553-9

This book is dedicated to Stacy, Carter and Maggie in memory of our life and times together in Grand Forks, North Dakota—"oh for fun"

Contents

Figures

Tables

Foreword

Judson Edwards is a successful practitioner and professor of economic development by vocation. He is a geographer at heart, and an acute (sometimes cynical) observer of local politics by nature. All these attributes and more are on display in this important book.

We're all hungry for success in economic development. Our jobs as practitioners depend, in large measure, on the short-term happiness of our local politicians and their constituents. As professionals, though, we're also dedicated to long-term improvements in the communities we serve and the well-being of the citizens who live there.

Economic development is a tricky business, though. We tend to jump on new ideas with fervor and confuse promising new tools with panaceas. Witness river valley development, enterprise zones, industrial clusters, creative communities more recently, and "going green" in the years ahead.

Telecommunications and other higher forms of technology fit a similar pattern: the sure salvation of struggling rural communities. Just plant a few miles of cable in the ground, like the newest genetic strain of seed corn, and good jobs will certainly sprout.

Edwards is a firm believer in the potential of technology for economic development, including in well-prepared rural areas. He reminds us in this book, though, that the technology tool—like any other in the economic development toolbox—can only work within a complex social, political, and geographic workshop.

Laying cable just isn't enough. Local workforces must be trained to capitalize on the opportunities that telecommunications infrastructure may make possible. Comprehensive development policies must encourage technology-based development. Geography is not dead; the success of wired communities remains largely a function of absolute and relative location, as well as access to highways and other traditional forms of infrastructure. In short, access to telecommunications technology may be a necessary condition for rural economic growth, but it is far from sufficient.

Finally, Judson is also encouraging. Put all the right pieces together in the right place, with the right people, and occasionally even thanks to the right politician—and all this technology stuff sometimes really works in a rural community that needs a boost. That's what it's all about; it's what keeps all of us looking for the right tool and the right way to use it.

Mark M. Miller, Ph.D.
Professor of Economic Development & Geography
The University of Southern Mississippi

Preface

My work in broadband telecommunications began through a visit to New Augusta, Mississippi—a struggling lumber community about 20 miles southeast of Hattiesburg, Mississippi along US Highway 98. I cannot remember from which mouth the words exited, but I can still hear the phrase, "once we get high-speed Internet in here things are going to start happening." This was in 1996, when "digital divide" was a common word and along with it came the promise that once the barrier to Internet accessibility was broken, economic success would soon follow. I began to study the promise of the "digital economy" for rural areas because of my curiosity as an economic geographer—understanding the key to rural economic engagement is the ability to defeat the distance/accessibility barrier. As the information economy emerged and accessibility in Mississippi began to grow, I awaited the stories of Mississippi miracles through telecommunications. Though there were a few successes in the state, the "miracles" did not overflow. At this point, I turned my focus to studying utilization not accessibility alone.

What can we do with "it" for economic development (with the "it" being broadband telecommunications)? This question was the driving force behind my work when it began in 1998, and ten years later, the same question rings true when working with rural communities throughout the United States. In 2002, when my dissertation research was completed, accessibility to broadband for rural communities was the focal point for policy-makers, local officials, and the economic development community—not the utilization of broadband. Though significant progress has been made in accessibility, the "it" question on ways to use the technology for creating economic development opportunities is still a serious issue in rural America.

As I try to assist communities in the development of broadband implementation strategies, I find myself reaching for my old research from 2002 and copying sections to distribute as part of the planning process. Though this research is 6 years old and some of the statistics dated, the information is still timely and holds validity. In order to spread the findings from this work, I decided to modify my original dissertation and publish it as a book. This book includes an update of the broadband policy research—which is a very important guide for developing strategic ideas for encouraging dissemination and implementation of broadband telecommunications.

In conclusion, I want the reader to know that I am not a "technology junkie"—but rather a "real world" academic dedicated to the economic development of rural communities. To me, broadband telecommunications is just one tool of many which can be used to help the economic growth of rural communities. Only when the integration of all community assets are enacted at the local level can greater economic opportunity be created for these remote communities.

Judson C. Edwards, Ph.D.
Associate Professor of Economics
Troy University

Chapter One

Introduction

> Deliverance: The act of setting free, the fact of being set free; liberation, release, rescue. The action of giving up, or of handing over or transferring; surrender.
>
> —Oxford English Dictionary 1993, 626

The world economic system evolved into an information-based, global economy over the past 20 years (Freeman and Louca 2002). The ability to participate and prosper within the "information economy" requires advanced information technology (IT) infrastructure, such as broadband telecommunications and computer technology, at every level of economic activity. Researchers suggest that the advent of advanced IT infrastructure allows virtually any location on the surface of the earth to actively participate in the global information economy (Kelly 1998; Cairncross 2001). For many remote areas, which were previously excluded from traditional economic activity because of their distance from urban centers of industry and trade, the information economy offers the needed opportunity for economic growth and development. Broadband technologies have proven to be a key component of rural communities' growth and prosperity, facilitating partnerships, creating online infrastructure, and expanding the market for goods and services. Although this technology has reached 99 percent of zip codes, these benefits are only beginning to emerge in rural America (U.S. House of Representatives 2007).

Even though advanced IT offers incredible possibilities, the achievement of economic sustainability and growth of rural communities in the United States has perplexed researchers and policymakers for decades (Rosenfeld 2004; Browne 2001; Rowley, Sears, and Nelson 1996). Traditionally, government-initiated rural economic development strategies focus upon "internal development issues, such as increasing educational opportunities, establishing agriculture/small-business loan programs, and improving physical infrastructure, because they are the only variables communities could enhance. Commonly, though, opportunities for economic growth were not available to most rural communities because of their physical location. Despite the efforts to improve "internal" conditions, the locational disadvantages of rural communities have been nearly impossible to overcome.

The capability of information technologies to make physical location less important has the potential to allow rural areas throughout the United States and the world to participate in the global economy without regard to geographic location (Turner 2002; Davis and Meyer 1998). Though economic development benefits are possible for remote areas in the information economy, research studies suggest that rural communities throughout the United States are struggling to adapt and apply information technologies for economic and social development. The United States Department of Commerce (1998) notes that no matter the income level, households in rural areas are less likely to own computers than households in urban or central city areas. Many economic development policy initiatives at the federal, state, and local levels encourage rural areas to invest in education, technology, and other elements to compete with urban areas or be left behind (Applied Rural Telecommunications Clearinghouse 1998; Carney 1996; Clark, Ilbery, and Berkerley 1995). As noted in the literature, the potential of IT must be realized or the socioeconomic divide between urban and rural areas will continue to grow (Allen and Dillman 1994; Cleveland 1995).

The rural-urban divide is driven by the inability of rural areas to support the needed advanced telecommunications infrastructure investment. Sparse population is considered a "rural penalty" due to the continuing concentration of hi-tech capabilities in large cities (Malecki 1997a, 125). A 1992 report on telecommunications and rural development in Nebraska notes that sparsely populated rural communities cannot possibly have all of the telecommunication support services and infrastructure available in urban areas because of low profitability associated with low population density. However, the same study suggests that rural communities should make investments to reposition themselves to participate fully in the new information-based economy, noting that "economic development of rural communities in the information age will demand a quality telecommunications facility" (Nazam 1992, 15). According to Edwin Parker (2000), rural communities without connection to emerging broadband networks will suffer the same economic fate as many communities that were passed by the telephone network, the railroad, and the interstate highway system.

Information technologies have the ability to increase the opportunities for rural areas in the economic development process, but sufficient broadband telecommunications infrastructure, defined as Digital Subscriber Line (DSL) or greater, is required if the potential benefits are to become reality. To assist in the technical explanation of high-speed telecommunications, Table 1-1 briefly summarizes the traits of "sufficient" residential and business broadband services.

The literature evaluates the benefits of private and public telecommunications technology investments and describes the power of regulatory and policy changes on private-sector business development (Malecki 1997b). In addition to policy evaluations, the research measures the impacts of telecommunications infrastructure both quantitatively and qualitatively in terms of jobs created, time to market reductions, and assessments of location factors (Haynes, Yilmaz, and Lall 1999). Despite these considerations, few studies attempt to produce a set of factors that identify rural communities with IT-based development potential and offer a set of practical guidelines for utilizing broadband telecommunications to promote rural economic development.

The main purposes of the research study are: 1) to determine the importance or lack thereof broadband telecommunications for rural economic development, 2) to develop a "broadband development index" for assessing the potential of communities to utilize broadband telecommunications for

Table 1-1. Residential and Business Broadband Types and Bandwidth Profiles

Broadband Service	*Speed*	*Description*
Residential		
Dial Up Service	56 Kbps	Basic Dial-up Modems - No comparison to Broadband Speed
Satellite	500 Kbps	Usually 150–500 Kbps downstream, 40–60 Kbps upstream.
DSL (ADSL)	1.544 Mbps	128Kbps to 1.5 Mbps downstream, 64Kbps to 1.5 Mbps upstream.
Cable Modem	2 Mbps	Usually 200Kbps to 2 Mbps upstream and downstream.
SDSL	2.048 Mbps	1.544/2.048 Mbps upstream and downstream.
Business		
Fractional T1 / ISDN	128 Kbps	128 Kbps to 1.544 Mbps Basic T1 speed with some of the 24/64 Kbps channels turned off.
Frame Relay	1.544 Mbps	64 Kbps–1.544 Mbps.
DSL (ADSL)	1.544 Mbps	128Kbps to 1.5 Mbps downstream. 64Kbps to 1.5 Mbps upstream.
T1, DS1 North America	1.544 Mbps	Point-to-point dedicated, digital circuit supporting data rates of 1.544 Mbps.

economic development, and 3) to offer policy recommendations for telecommunication-based economic development planning strategies.

This book begins carrying out these objectives with two necessary steps. These steps are the identification of rural communities with broadband telecommunications accessibility and the discovery of the most influential factors contributing to or deterring the utilization of broadband telecommunications for rural economic development. The conclusion of this section explains the complete research methodology by providing the procedural details of each chapter analysis. Results from the research chapters provide a foundation for the construction of a model for rural economic development planning through the application of broadband telecommunications.

BACKGROUND TO THE RESEARCH PROBLEM

Rural Economic Development in the United States: History, Policy, and Future

Simply put, rural areas are in serious trouble. Just as the title of this book suggests, the majority of the American rural communities are in need of deliverance from their economic tribulations. The movement of the national economic structure from manufacturing to a service-based economy, the migration of people away from rural communities to urban centers, high unemployment rates, and the lack of modern telecommunications infrastructure are just a sample of the numerous obstacles facing rural communities throughout the United States. In order to maintain the rural population, there must be a way to make a living. For example, based on statistics from the Alabama State Data Center, urban areas in Alabama gained 10,400 jobs from May 2007 to May 2008. During the same period all other counties lost 1,900 jobs. Accordingly, the most important issue for rural America is the achievement and sustainability of economic development.

It is elementary to recognize the limited economic opportunities in rural communities when compared to urban areas. As easy to understand is the low population density that contributes to the lack of economic sustainability in rural areas. So, if the rural areas of the United States are predetermined to fall behind urban areas in economic development opportunities, why should we care if they survive? The answer is quite simple; rural America provides essential economic and cultural functions that support both urban centers and nation as a whole (Ryan 1990).

Rural America supports the economic functions of urban areas through such contributions as surplus labor support and markets for urban goods and services. Traditionally, labor movement from rural areas reinforces the growth capacity within industries and businesses in urban centers. Despite the increased employment opportunities available in urban areas, many workers in rural America are not perfectly mobile and remain. The literature suggests that rural workers are often more attached to their communities than urban workers, therefore less geographically mobile. This attachment to place drives the demand for rural economic development activities, which commonly results in the hiring of a local economic developer.

In addition to economic contributions, another significant contribution of rural America is cultural. The attachment of Americans to the land is a cultural foundation of the United States. Although much diversity exists within rural regions of the U.S., i.e., the Mississippi Delta and Midwest farming towns, there is an idealistic sense in which rural America constitutes a unity (Galston and Baehler 1995). According to Castle (1995), the cultural identity of the United States rests upon its rural heritage. Rural communities stand for something more than just a place; they represent the unique dichotomy between the openness of its landscape and closeness of its people. From a sociological perspective, a person's experience with a place is both a serious environmental issue and a deeply personal one. According to Hiss (1990, 13), people relate to place "with a close bond, intricate in nature, and not abstract, not remote at all: It's enveloping, almost a continuum with all we are and think."

Historically the economic survival of rural America has been a continuous struggle. The history of economic development in rural America is a roller coaster ride of successes and failures. Although the rural economic history of the United States can be traced to the pre-colonial period, the past three decades sufficiently illustrate the up and down nature of the rural economy. These rural development periods are: 1) the "Rural Renaissance" of the 1970s, 2) the "Decade of Decline" of the 1980s, and 3) the "Deindustrialization Decade" of the 1990s.

Rural America experienced a "rural renaissance" during the 1970s. According to Flora and Christianson (1991), the "rural turnaround" was initiated by the expansion of world purchasing power (triggering price increases for all basic commodities), and favorable monetary and fiscal policy. The low value of the U.S. dollar made exports expensive, so industries began expanding operations into rural areas because of the cheap labor, inexpensive land, and relaxed environmental regulations (Bradshaw and Blakely 1999). Thanks to the growth of extractive industries and routine manufacturing, many rural communities became "industry towns," depending on a single company for the majority of local jobs. The "industry town" structure established during the 1970s would prove costly during the following decades.

In the 1980s, rural Americans suffered one of the longest, most severe economic declines in several decades (Parker, et al. 1989). In the "decade of decline," out-migration from rural areas to cities reached rates that exceeded that of the "golden age" of industrial development during 1950s and 1960s (Johansen 1993). Galston and Baehler (1995) survey and examine the rural economic decline of the 1980s through the following socio-economic indicators: employment, unemployment, income, wages, poverty, and population. These socio-economic indicators, presented in Figure 1-1, compare the rural and urban economies of the 1980s.

Also of consequence to the decline of rural areas was the farm crisis of the 1980s. The crisis resulted in large numbers of farmers leaving their property for employment in industry within larger towns and metropolitan regions. Toward the close of the decade, research literature began to address the economic transformation of rural areas, moving from dependence upon agriculture and industry toward the service industry. This rural transformation is referred to as "deindustrialization" (Falk and Lyson 1991, 12).

The 1990s represent a decade of structural economic change for rural communities in the United States. Rural areas began to embrace alternative strategies for economic development, such as tourism and retiree attraction (Berryhill 2000). The realization that technology decreased the need for employees in manufacturing and agricultural production triggered rural commu-

Employment: Between 1979 and 1987, metropolitan area employment grew by 18 percent, while rural employment grew by only 8 percent

Unemployment: Between 1979 and 1987, annual unemployment ranged between 1 percent and 2.5 percent higher than metro rates

Income: The ratio of rural to metro per capita incomes declined from 77 percent at the end of the 1970s to only 73 percent in 1987

Wages: After adjusting for inflation, average annual earning per job fell by 8 percent ($1700) in rural areas between 1979 and 1987, compared to 2 percent ($450) for metro areas. As a result, the metro/urban gap grew from $5000 to nearly $6200.

Poverty: The rural poverty rate soared to 18 percent between 1979 and 1982, and remained at that level throughout the decade. By the late 1980s, the rural poverty rate was nearly 50 percent higher than metro rates.

Population: While the rural population growth rate had exceeded the metro rate by almost 40 percent in the 1970s, it fell to less than half the metro rate through the 1980s. By the mid-1980s, annual out-migration from rural areas reached nearly 500,000, a rate substantially above the annual average for the 1950s and 1960s. More than one half of all rural counties lost population during the 1980s.

Figure 1-1. Selected Indicators of Rural Decline in the 1980s. *Source:* Galston and Baehler 1995; Deavers 1989.

nities to focus recruiting efforts on service industries. Service industries, such as call centers and back office operations, began taking advantage of low-cost labor in rural communities, thanks in part to advances in telecommunications and information technologies. The "deindustrialization" of the 1990s forced rural communities to conduct economic development efforts from a stance of self-evaluation and self-improvement. The emerging economic development strategies included strategic planning, human resource advancements through increased education, and assessments of community infrastructure (Plugge 1996; West 1999).

The rise and fall of rural communities over the past 30 years suggests that remote areas of the United States are vulnerable to affects of global economic trends and macro-level policy decisions, factors rural communities have little ability to influence. However, federal and state governments do have some capacity to influence and encourage the economic development of rural areas through strategic policies.

Finding ways to stimulate economic development in rural America is a continuous topic of concern for policymakers at all levels of government. But, policymakers are not alone in their attempts to improve opportunities for economic growth. Academic research offers numerous economic development theories and models from various academic fields of study, yet few produce meaningful results for practical application. This failure to develop successful models and policies for economic development in rural America is well documented. The models and strategies of the past—large federal public investment programs, industrial attraction campaigns, and resource exploitation strategies—generally produced disappointing results (Galston and Baehler 1995). It is noted by Bradshaw and Blakely (1979, 63) that theories of economic development are "ill-equipped" to explain many of the economic and social problems facing rural areas. If history is any guide, the potential of policymakers or researchers to offer successful strategies for assistance is less than optimistic.

Historically, the rural economic development policies of the federal government focused primarily on agriculture, and can be divided into three distinct eras (Castle 1993). According to Pulver, distinct differences exist between the three periods:

> During the first era, rural policy was focused on overcoming space, encouraging settlement, and creating opportunity in rural areas. Mail service for rural areas, rural roads and the initiation of Land Grant Universities were all aspects of rural policy in that era. The second era, which began at the turn of the century, emphasized managing forests for long term economic use, the development of water for purposes of economic opportunity and the creation of university research and extension programs aimed at improving rural living conditions and farm

> productivity. The third era, which started in the depths of the Great Depression, saw the beginning of public support for rural electrification, soil conservation, farm credit, and agricultural price supports (Pulver 1998, 3).

The belief that if farms flourished all of rural America would experience economic development has not proven itself true. The importance of farm employment has steadily fallen since turn of the twentieth century. In 1910, two-thirds of the people in rural America lived on farms. Yet in 1990, only one in sixteen people lived on a farm (Hart 1992). Despite the shrinking of farm population, rural policy still reflects the bias toward farming and agricultural policy. Policy initiatives continue to focus on farm product prices, the export of agricultural products, the utilization of farm products in poverty related programs, farm land conservation, and increased farm productivity (Pulver 1998).

The majority of rural policy is formed by the United States Department of Agriculture (USDA) due to their direct oversight of programs dealing with American farmers. Though the Economic Development Administration (EDA) is generally recognized as the federal arm responsible for economic development assistance, the USDA operates the rural economic development program. As a result of this structure, the political support of the USDA and its policies has fallen almost exclusively to farm organizations and agricultural commodity groups (Berryhill 2000). Many other important policy issues, such as education, telecommunications, transportation, and welfare are seen as the responsibility of other federal agencies. Due to USDA oversight of rural policy development, there has been little reflection regarding the unique and special concerns of rural regions relative to issues beyond farm income and productivity (Pulver 1998).

Local leaders and economic development practitioners clearly recognize the lack of innovative federal policies to address the non-agricultural concerns of rural citizens, such as non-farm employment and telecommunications. According to Milan Wall, Co-Director of the Heartland Center for Leadership Development, federal policy must move away from its agricultural bias and become more innovative in rural economic diversification:

> I don't want to say anything anti-agriculture, but much of the federal policy really is geared toward agricultural production, and very little of it toward supporting broader diversity in rural economies. We need to have successful farmers, but we also need to recognize that it takes fewer and fewer people to till the same acres. If we are going to keep people in rural communities, it means we have to have a bigger supply of jobs that aren't directly in agricultural production (Bird 2001, 3-4).

Likewise, Neil and Tykkylainen (1998) note that aside from agriculture, industry-specific programs for rural communities that address structural changes in community and area development are not part of federal policy. While the majority of federal rural policy continues to focus on agricultural-related issues, state policy has traditionally compensated for federal policy deficiencies through being more responsive in the creation of programs and initiatives to address the economic development needs of its rural citizens and communities (Johnson 2001).

The economic development literature thoroughly documents both the effectiveness and content of state economic development policies (Bartik 1991; Goss and Phillips 1997; Accordino 1994; Bradshaw and Blakely 1999; Langer 2001). The role of state involvement in the economic development of rural areas is reflective of the structural changes in the economic system. These structural adjustments include the transformation of rural employment from agriculture to manufacturing and manufacturing to service/information. The literature identifies changes in state policies as the "three waves" of state economic development policy. Figure 1-2 presents these economic development waves.

The first wave of state economic development policy is industrial attraction efforts. The economic development studies refer to these industrial attraction

FIRST WAVE:	**Industrial Attraction Efforts** • Dominated by programs designed to attract footloose firms from old industrial regions to growing regions • Typical tools: subsidized loans, direct payments, tax reductions, competitive and expensive industrial recruitment programs
SECOND WAVE:	**Industrial Retention Efforts** • Shift focus from attracting firms from out of state to retaining and expanding existing firms • Typical tools: creating new businesses, increasing capital investment, developing incubators, technical assistance
THIRD WAVE:	**Industrial Policy Orientation** • The new programs reflect community and business collaborations designed to build entire industrial sectors that are globally competitive. • More policy oriented than project oriented • Typical tools: public-private partnerships, strategic planning, focus on cluster development, network building, leadership development, global competitiveness

Figure 1-2. Three Waves of State Economic Development Efforts. *Source:* Bradshaw and Blakely 1999.

efforts as "smokestack chasing" (Swager 1991, 28). The notion of recruiting industry from traditional manufacturing centers to growth regions, such as the Sunbelt, was a reflection of the changing nature of industrial location toward areas of available, low-cost labor. This increasing availability of labor for industrial employment was a clear indication of the declining importance of agriculture in many of the newly industrialized states, such as Alabama, Mississippi, and Georgia.

The second wave moved the focus from out-of-state industrial recruitment to in-state retention and expansion programs. This second wave recognized existing industry as the most important component to economic development sustainability. Clarke and Gaile (1992) note that second-wave strategies are typified by a strong investment and entrepreneurial approach, including revolving loan funds, below-market loans, enterprise zones, and tax increment financing. These second wave strategies also include the establishment of existing industry-monitoring programs within state economic development departments, such as the Mississippi Development Authority and Georgia Department of Industry, Trade, and Tourism.

The third wave of state economic development efforts is industrial policy. According to Bradshaw and Blakely (1999), third-wave strategies shift the focus of local development by creating the context for economic growth through public-private partnerships, networks that leverage capital, and human resources to increase the global competitiveness of strategically-linked firms. The state level policy within the third wave focused on positioning the state to compete in a global environment. According to Foster:

> States are now more concerned with the overall performance of the state economy in achieving high levels of productivity and competitiveness that increase income and provide a high standard of living and quality of life for all residents. Instead of simply creating jobs, they seek high-skill jobs. Instead of simply promoting businesses that will create jobs, they are more interested in generating high performance businesses that are competitive in quality, attractive to customer needs, and timeliness, as well as costs. They are concerned with the ways in which workers and businesses interact in networks and clusters. And they are interested in the dynamics among those economic entities and related social and political institutions within the context of specific regions and communities (Foster 1992, 5).

Another important aspect of third-wave development is the move away from "free gift" programs, such as direct payments to industries, found within the first wave, to internally focused programs, such as strategic planning and cluster development.

The purpose of evaluating history is to learn from past experiences and improve the promise of the future. The three waves of state economic development efforts reflect important changes in the economic structure of both the domestic and global economy. The economic development waves also mirror the past thirty years of rural economic development in the United States, from the industrialization of rural areas in the 1970s, with the first wave efforts, to strategic planning in rural communities in the 1990s, with third wave industrial policy.

Examining the economic and policy history of the rural United States helps answer the question: What is the economic future of rural communities in the 21st century?

To answer this question, it is necessary to understand the economic conditions in which rural communities must compete. The economic setting of the present and future is commonly referred to as the "new economy." The new economy can be defined in several ways. One of the earliest definitions of the new economy came from Shepard:

> By the new economy, we mean two broad trends that have been underway for the past several years. The first is the globalization of business. Simply put, capitalism is spreading around the world—if not full blown capitalism, at least the introduction of market forces, freer trade, and wide spread deregulation . . . For the United States, this means international trade and investment play a much greater role in our economic life than ever before . . . The second trend is the revolution in information technology. This one is all around us—fax machines, cellular phones, personal computers, modems, and the Internet. But it's more than that. It's the digitization of all information—words, pictures, data, and so on . . . Housing and auto used to drive the U.S. economy, now information technology accounts for a quarter to a third of economic growth . . . Furthermore, information technology affects every other industry. It boosts productivity, reduces costs, cuts inventory, and facilitates electronic commerce. It is, in short, a transcendent technology—like railroads in the 19th century and automobiles in the 20th (Federal Reserve Bank of Atlanta 2000, 11).

Thus, according to Shepard, the new economy is composed of two major components: globalization and information technologies. But, perhaps more important is the recognition of the new economy as being "transcendent" in nature; not only producing something new, but also changing the structure of existing businesses and industries. An example is the advent of e-commerce. It is defined as the process of two or more parties making business transactions via computer and some type of network (Phillips 2000). For example, the use of e-commerce is common within the production processes of automobile manufacturing plants (Mayer and Greenberg 2001).

The "transcendent" nature of the new economy can be better understood when comparing it to the "old" economy. The old economy was built on a manufacturing base geared toward standardized production and operated by stable, hierarchical organizations focused primarily on a national marketplace and preoccupied with incremental cost reductions through economies of scale and scope (Cohen, DeLong, and Zysman 2000). In comparison, the new economy is knowledge and idea-based, where keys to wealth and job creation are the extent to which ideas, innovation, and technology are embedded in all sectors of the economy—services, manufacturing, and agriculture (Atkinson 2000). In order to further clarify the differences between the old and new economy, Table 1-2 summarizes both in terms of economy-wide characteristics, industrial structure, workforce, and government policy.

Table 1-2. Selected Issues in the Comparison of the Old and New Economy

Issue	*Old Economy*	*New Economy*
Economy-Wide Characteristics		
Markets	Stable	Dynamic
Scope of Competition	National	Global
Organizational Form	Hierarchical	Networked
Potential Geographic Mobility of Business	Low	High
Competition Between Regions	Low	High
Industrial Structure		
Method of Production	Mass	Flexible
Key Factor of Production	Capital/Labor	Knowledge
Key Technology Driver	Mechanization	Digitization
Source of Competitive Advantage	Economies of Scale	Innovation
Importance of Research	Moderate	High
Relations with other firms	Go It Alone	Collaboration
Workforce		
Principal Policy Goal	Full Employment	Higher Income
Skills	Job Specific	Broad Skills
Requisite Education	A Skill	Lifelong Learning
Labor-Management Relations	Adversarial	Collaborative
Nature of Employment	Stable	Risk/Opportunity
Government Policy		
Business-Gov't Relations	Impose Requirements	Assist firms' Innovation /Growth
Regulation	Command and Control	Market Flexibility

Thus, based on the characteristics of the new economy, is rural America poised to compete and prosper within its environment? The research is varied. According to the economic development literature, there are three key obstacles to rural participation within the "new" economy: 1) poor technological infrastructure, 2) lack of innovation and entrepreneurialism, and 3) inability to compete with urban centers (Beck 2001; Davis and Meyer 1998; Allen and Dillman 1994).

Modern technological infrastructure is critical for participation within the new economy. Unfortunately for rural America, there are numerous physical and economic barriers for inclusion in the new economy. The problems associated with people and businesses without access to basic information technologies are commonly referred to as the "digital divide." Although the "digital divide" is comprised of numerous components, such as low rates of computer ownership, one of the major pieces of the "digital divide" is the lack of broadband Internet access to rural communities throughout the United States. A federal study of the "digital divide" notes that if rural areas of the United States are "without affordable broadband Internet access, existing rural businesses will be denied the opportunity to participate in the digital economy" (White House 2000, 8). Among the reasons for the lack of deployment to remote areas is low population density, meaning potential customers are few and widely dispersed; therefore, rural areas do not represent an attractive market for private investment (Beck 2000).

A second major obstacle to rural prosperity within the new economy is the lack of innovation and entrepreneurship. For many rural communities, the economic and social structure associated with the farming or "industry" town is not easily changed. The "assembly line" mentality of many manufacturing-based rural areas does not lend itself to the development of innovative ideas and new technology-based businesses. The literature recognizes that rural states are not well prepared for the new types of economic competition that place emphasis upon process. According to Mitchelson: "what meager inventiveness in broadband telecommunications infrastructure or physical innovativeness is found within the urbanized areas of the states" (Mitchelson 1999, 218).

A third major obstacle to rural success within the new economy is the inability to compete with urban centers. The evolution of local economies directly depends on the development of localized labor skills and the agglomeration of linkages among local firms (Malecki 1997a). Many urban centers have prospered immensely in the new economy because of their ability to take advantage of the agglomeration of similar high-tech industries, research facilities, and venture capital firms. In general, rural areas do not have the social or physical capital to compete with urban centers that continually build upon their agglomeration capacity.

The underlying message of this research and commentary is that rural areas of the United States must begin to prepare for participation in the information economy. Although there are many components to the new economy, the emergence of broadband technology has widespread implications for the development of rural America. In order to share in the fruits of the new economy, rural areas must make capital investments in broadband infrastructure (Beck 2000). Due to investments in broadband telecommunications, many rural communities have experienced both economic and social benefits through direct job creation, increased educational opportunities, improved healthcare, and increasing market size for local businesses (Maturi 2001). As noted in Table 1-2, the macro-economic shift from a manufacturing to service-based economy demands that rural areas adopt innovative strategies and technologies in order to actively participate in the new economy.

STATEMENT OF PURPOSE

The purpose of the book is to examine the potential role of broadband telecommunications in the economic development of rural communities in the United States and develop a structural framework for guiding broadband-based rural economic development planning. The book is structured to ascertain the following: 1) evaluate the importance of broadband telecommunications to the economic development of rural communities, 2) develop a "broadband development index" for assessing the potential of rural communities to utilize broadband telecommunications for economic development, and 3) determine the role of leadership and policy in telecommunications-based economic development planning.

In practice, the research results focus on "real-world" use by economic developers for the improvement of rural communities. Thus, a secondary research purpose is to report the findings of the research in a manner that appeases the scrutiny of the academy, yet so fundamental that any economic development professional can easily put into action the results. This secondary purpose is a necessity to achieve the purpose of the research.

CONCEPTUAL AND THEORETICAL FRAMEWORK

The theoretical foundation of this research is cluster theory. Cluster theory is one of the latest strategies used to guide and inform technology policy, yet despite its numerous applications, there is no standard methodology (Stough et al. 2000). Although cluster-based methodologies are quite diverse, they focus on four broad attributes that shape the competitive environment in which local firms compete:

1) *Factor conditions*-the nation's position in factors of production, such as skilled labor or infrastructure, necessary to compete in a given industry;
2) *Demand conditions*-the nature of home demand for the industry's product or service;
3) *Related and supporting industries*-the presence or absence of supplier and related support industry; and
4) *Firm strategy, structure, and rivalry*-the conditions in the nation governing how companies are created, organized, and managed.

The relationships between these four national system determinants of competitive advantage are referred to as the "diamond" (Porter 1990). Clusters are defined by the interaction of these determinants, which as a system drive the competitive advantage of a location. Figure 1-3 presents the "diamond" of competitive advantage.

In addition to the "diamond" determinants, two supplementary variables are chance and government. The economic structures of rural communities

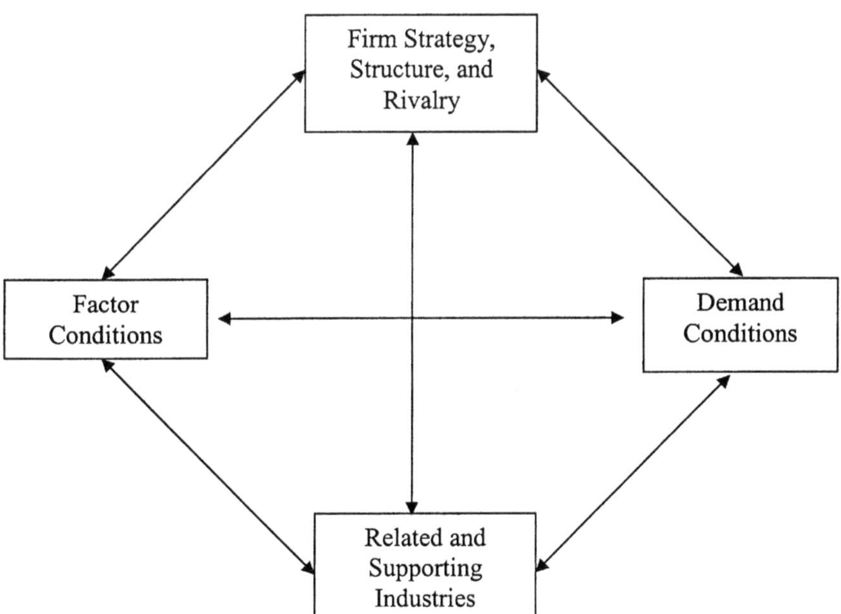

Figure 1-3. "Diamond Model" of Cluster Theory. ***Source:*** **Porter 1990.**

are not as complex as the national economies presented in the foundational work; therefore, all of the determinants of the "diamond" may not be applicable. Porter (1990, 72) recognizes that less complex economic systems could develop competitive advantage with only two of the four factors present.

This research employs cluster theory because it provides a foundation from which to determine the economic development capacity of rural communities within the framework of the new economy. Evans and Miller (2000) note that clusters provide the capacity and flexibility to react quickly to innovations. This study concentrates on four of the six components of cluster theory: factor conditions, demand conditions, government, and chance. Chapter Four offers specifics about the application of cluster theory in this research.

DEFINITION OF TERMS

It is important to establish the foundation for many of the ideas presented throughout the various chapters. Words such as "rural" and "broadband" can be defined in numerous ways. In accordance, the following terms are significant to the research, thus it is necessary to offer definitions and establish these meanings. These fundamental vocabularies are:

Rural—cities or census-designated places with a population of less than 2,500 and counties of less than 15 persons per square mile (US Census Bureau 1995).

Broadband telecommunications—the transmission of voice, video, and data at the same time over fiber optic telephone lines; broadband ISDN can support data rates of 1.5 million bits per second (bps) (Webopedia 2001).

Economic development planning—undertaken for the purposes of increasing standards of living through greater per capita income, quality and quantity of employment opportunities, and in general, increased quality of life for persons within the designated economic development region (Phillips 1996).

Centers of Excellence in Rural America (CERA)—CERA is an effort to test the hypothesis that creating a network of small rural towns deploying affordable, high-speed telecommunications services will result in increased job creation and/or income in those towns while also improving access to education, healthcare, and governmental services. CERA is a multi-state, multi-site project sponsored by the Western Governors' Association, with leadership from the governors of North Dakota and Wyoming and Watford City and Mayville in North Dakota (Western Governors' Association 1997).

STUDY AREAS

The first important step in conducting this research is the identification of locations that serve as the focus of study. In accordance, the rural counties of North Dakota, Minnesota, Montana, and Nebraska are the areas of study. The rural counties are selected for this research because: 1) the upper Midwest is one of the most rural regions of the U.S.; and 2) it contains the communities involved in the CERAtown program.

The purpose of the CERAtown program is to test the hypothesis that creating a network of small rural towns deploying affordable, high-speed telecommunications services will result in increased job creation. With the significant assumption being that these technologies will alleviate the distance barrier experienced by their physical location. The CERAtown counties examined in this research are: Traill County, North Dakota; McKenzie County, North Dakota; Niobrara County, Wyoming; Park County, Wyoming; and Converse County, Wyoming.

Another significant study area consideration is the ability to compare rural communities within the same region that are without affiliation with the CERAtown program. Thus, the rural counties of North Dakota and Wyoming are selected because they are the two states that contain the member communities of the CERAtown project and provide a useful comparison to the CERAtown counties.

A third important reason for the selection of the study area communities is having a large enough sample size to conduct the factor analysis. Thus, the rural border counties of Montana, South Dakota, and Nebraska are included in the study because they most closely resemble the counties of the CERAtown states of North Dakota and Wyoming, both in economic structure and geography. In total, the study area consists of 96 counties with the following distribution: 45 in North Dakota, 20 in Wyoming, and 31 in Montana, Minnesota, Nebraska, and South Dakota. Figure 1-4 presents a map of the study areas. Further details of the research methodology are presented in Chapter Four.

Extending beyond the quantitative analysis, LaGrange, Georgia, USA and Ennis, Ireland, U.K. are included as case studies within the qualitative portion of the research. These communities are used to define "success" with regard to broadband telecommunications-based economic development planning. These places represent "success" because of their external awards and recognition in academic literature.

In 2000, LaGrange, Georgia won the "Most Intelligent City Award" given by the World Teleport Association (WTA), which consists of members from 20 member nations. The award is given yearly to recognize a community

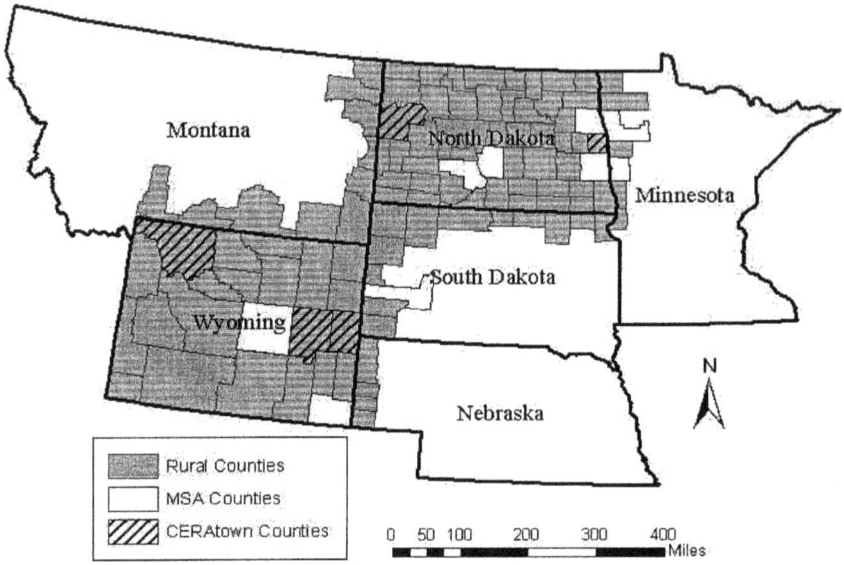

Figure 1-4. Areas of Study.

with a documented strategy for creating a "digital age" economy using information and communications technology. The WTA award is for communities that emphasize broadband application, attract leading-edge businesses, stimulate job creation, and generate economic growth. LaGrange, Georgia was the first city in the United States to win the award. The academic literature also recognizes LaGrange as successful (Youtie 2000, Harvard Business School 2001).

Ennis, Ireland, "The Information Age Town" is recognized internationally as a model for implementing information technology for economic development. The city received a 19 million (US$ 18.5 million) award from *eircom*, Ireland's largest telecommunications company. With the funding, Ennis, Ireland became the test location to study what happens when the advanced information technology is introduced to an entire community on a large, cost-reduced scale. According to David Byrne, EU Commissioner of Health and Consumer Protection, "Ennis is a pioneer town . . . and is being held up to other countries as an example of how to achieve a dynamic economy" (*eircom* 1996, 11). A comprehensive study from the Ennis project revealed that 83% of all households own an Internet-enabled computer and confirms the community as a "best practice" model (Industry Ennis News 2001, 4).

OUTLINE OF CHAPTERS

This book consists of six chapters. Chapters One, Two, and Three serve as the foundation on which the original research is built. The three chapters provide a background to the rural economic development in the United States; establish the importance of broadband telecommunications accessibility for rural economic development; and review the role of policy in the deployment of broadband telecommunications. Chapter Four applies cluster theory for rural economic development, while Chapter Five focuses on the case study CERA-town communities. These chapters utilize both quantitative and qualitative techniques to build a "structural" model for broadband telecommunications-based rural economic development planning.

The following outline summarizes the chapter content and presents the focal research question for each.

Chapter Two, *Broadband Telecommunications: The Interstate Highway of the 21st Century*, addresses the following research question: what are the necessary technological conditions for economic development in the information economy?

The purpose of the chapter is to determine the importance of broadband telecommunications for economic inclusion in the 21st Century information economy. The research methodologies for this section consist of the following: review of literature, presentation of descriptive statistics relating to broadband telecommunications and economic development, and developing maps through Geographic Information Systems (GIS) to provide a spatial dimension to the descriptive statistics. The review of literature serves as the introduction to the chapter and establishes a foundation for determining the importance of broadband for economic development in the information economy. The descriptive statistics and GIS are combined to present data on the relationship between economic, social, and spatial aspects of broadband technology and development.

Chapter Three, *Federal and State Broadband Policies: Shaping the Prospect of Economic Development in Rural America*, focuses on the following research question: what are the proposed policy initiatives at both the federal and state levels with regard to broadband telecommunications?

The purpose of this chapter is to identify innovative federal and state policies in the deployment of broadband telecommunications in the United States. If broadband telecommunications is to be examined as a possible tool for encouraging economic development in rural communities, it is important to understand the policies that support or deter its access. This section provides an important foundation for the remainder of the research and contributes to the formulation of a model for rural economic development planning based upon broadband telecommunications. The data used in the chapter is derived from the Congressional Record of the 106th and 107th U.S. Congress and a fifty-state survey of legislative and executive proposals.

Chapter Four, *Beyond Access: A Cluster Approach to Broadband Telecommunication-Based Rural Economic Development Planning*, addresses the following research question: beyond access, what factors are important for rural economic development based upon information technologies and broadband telecommunications?

The chapter provides the theoretical foundation of the research through the examination of cluster theory. Factor analysis is used to determine which socio-economic variables, based upon existing literature, are the most significant in the application of broadband telecommunications for rural economic development within the study area counties. The results of the factor analysis are used to determine which of the CERAtown communities are most likely to be "successful" in implementing the technology for economic development by ranking the data based on the emergent factors with the "broadband development index." Also, the analysis identifies the potentially "least successful" CERAtown community. Of the five communities, two are selected as "best case" and "worst case" communities for use in the case study research of Chapter Five.

Chapter Five, *Local Leadership and Policy in Broadband Telecommunication-Based Rural Economic Development Planning*, focuses on the following research question: what is the importance of local leadership and policy in the creation of a rural economic development model based upon broadband telecommunications?

The chapter determines the impact of leadership and policy in IT-based cluster development. This section also includes the culminating product of this study, a structural model for broadband telecommunications-based rural economic development planning. The research question is answered through literature review and multiple case study research. The qualitative portion of the chapter seeks to address the two supplemental categories of the competitive advantage theory, chance and government/policy through structured, in-person interviews. The interviews provide additional, non-statistical information to complete the cluster analysis evaluation, as well as, determining the importance of leadership and policy in broadband application/clustering success. Two "successful" case study communities are introduced, Ennis, Ireland and Lagrange, Georgia and used as a benchmark against the selected CERAtown communities. Interviews are conducted in the "best case/worst case" CERAtowns identified in Chapter Four. The following community leaders are interviewed in both the "successful" communities and selected

CERAtowns: small-business representative; agricultural/service sector representative; project representative; education representative; and economic development representative. The chapter concludes with a model for rural economic development planning based upon research results from the previous chapters. The model includes policy recommendations, community evaluations based upon cluster analysis, and additional planning strategies revealed through the qualitative research.

The brief final chapter, *Conclusion and Recommendations*, summarizes the findings of the study and offers recommendations for future research.

Chapter Two

Broadband Telecommunications: The Interstate Highway of the 21st Century

Throughout history, the economic fate of U.S. cities and communities has depended largely upon their accessibility. A hundred years ago, the lack of a railroad stop condemned many towns to a lingering death (Fidelman 1994). Forty years ago, the interstate highway system helped communities with interchange access prosper, while those on the back roads stagnated. Traditionally, the ability to move goods to various production facilities and markets were dependent upon efficient transportation systems, such as water, rail, and highways. These "traditional" location factors were critical in determining where businesses dependent upon raw materials and physical movement of goods would establish operations.

But unlike the long-established movement of physical commodities along transportation networks, the "new economy" is concerned with the movement of information. The "new economy" operates in two distinctive systems, one based upon tangible, physical items, and the other based upon intangible, ideas, and information. Moschella (1997, 18) distinguishes the "new economy" as dualistic in nature, with traditional industries, such as manufacturing and retail, based on "atoms" and information industries, such as banking and services, based on digital "bits." Just as traditional manufacturing plants create products based upon "atoms," "bits" of information created within information-based industries do not appear magically. Skilled people develop them. Although "traditional" location factors are still important to certain industries, companies involved in the movement of information, such as service industries, are increasingly dependent on the ability to link locations with skilled human resources (Kotkin 2000). The ability of a company to connect to skilled labor and transmit "bits" of information depends upon telecommunications infrastructure for connectivity.

Based on the need to link locations rich in human resource capacity, a key factor for development in the "new" economy is connectivity. So just as the movement of physical goods along railroads and interstate highways played important roles within the structure of the "old" economy, telecommunications appears to be the critical linking mechanism within the "new" economy. According to Former U.S. Secretary of Commerce, William Daley, "We need to encourage the building of a broadband infrastructure that allows all Americans to have access to the advanced services that support the Internet . . . to realize the full potential of this digital economy, every person and every business must be able to participate fully and make their own unique contribution to its development" (U.S. Department of Commerce 1999).

The purpose of this chapter is to explore the question, what are the necessary technological conditions for economic development in the information economy? The research hypothesis is that broadband telecommunications are necessary for economic development in the information economy. The examination of the potential role of broadband telecommunications for economic development in the information economy consists of the following: 1) an overview of broadband telecommunications, 2) establish the importance of broadband telecommunications in the new economy, and 3) determine the significance of broadband connectivity in the economic development of rural communities.

BROADBAND TELECOMMUNICATIONS: A TECHNICAL AND SPATIAL OVERVIEW

Advanced telecommunications capability allows users to send and receive large amounts of information. More specifically, with advanced telecommunication capability consumers can take advantage of advanced services that allow residential and business customers to create and access content, sophisticated applications, and high-bandwidth services. For example, advanced services allow businesses and their customers quickly to exchange data over long distances, doctors to provide real-time diagnosis to patients in remote areas, people with hearing and speech disabilities to communicate through video links using sign language, teachers to create interactive multimedia learning environments for their students, and individuals to have faster, more robust access to the Internet.

Broadband is a term used to describe advanced telecommunications systems with high data transmission rates. It is defined by the Federal Communications Commission (2001) as "the capability of supporting at least 200 kilobits/second in the consumer's connection to the network, both from the provider to the consumer (downstream) and from the consumer to the provider (upstream)." Although broadband telecommunications is commonly referred to as "high-speed" or "advanced telecommunications," slightly different definitions exist. Advanced telecommunications is defined as "high-speed, switched, broadband telecommunications capability that enables users to originate and receive high-quality voice, data, graphics, and video telecommunications using any technology" (FCC 2000). A more general characterization of broadband telecommunications is a connection that is on all the time, everyday, and delivers Internet speeds generally ten times that of traditional dial-up Internet service. Broadband telecommunications are delivered in three basic forms: cable modem, digital subscriber lines (DSL), and fixed wireless.

Cable modem service is a commonly used connection for residential service and home-based businesses. Because most homes in the United States subscribe to cable television, high-speed Internet service is potentially as easy as installing a cable ready modem. Many will see this type of service as the easiest and the most economical entry into homes and small businesses (National Center for Small Communities 1999; McKinsey Quarterly 2001). According to Beck (2001), approximately 50 million homes and small businesses with Internet access are capable of receiving high-speed access through cable-based service. Another unique feature of cable modem service is that a personal computer (PC) is not a requirement to utilize the Internet. For example, in LaGrange, Georgia, citizens who subscribe to the locally operated cable company have free access to high-speed Internet through an interface device offered by WorldGate Communications. Despite the unique opportunities of cable modem service, a potential drawback is the inevitable reduction in service speed as the number of cable channels offered to consumers increases (Harvard Business School 2001).

DSL, the most common form of broadband service, is a digital connection over existing telephone service. The connection speed of the Internet service with DSL is dependent upon type of delivery line. Fiber optic lines carry the highest capacity service, yet forms of DSL are available over modified copper phone lines. Speed of service is also dependent upon distance from the central switching facility and the type of local switching service available. The most common varieties of DSL are ISDN (Integrated Service Digital Networks), ADSL (Asymmetric Digital Subscriber Line), and SDSL (Single or Symmetric Digital Subscriber Line). ISDN allows users to receive both voice and data calls on the same line, where users can browse the Internet and talk on the phone at the same time. ADSL moves data over existing copper phone lines at one-way speeds of up to seven megabits per second. SDSL moves data over single copper lines in both directions at the same speed, unlike ADSL that utilizes pairs of lines (dslreports.com 2002).

Beyond the copper-line based systems are the fiber-optic systems known as advanced DSL broadband telecommunications. Advanced DSL broadband telecommunications is divided into three categories, based upon speed: T1 (Digital Signal Level 1), T3 (Digital Signal Level 3), and OC3 (Optical Carrier 3). T1 lines transfer data at 1.5 megabits per second. T3 lines are capable of transferring data at 45 megabits per second, which is approximately 100 times faster than traditional dial-up 33–56K modems. OC3 carries data at 155 megabits per second over dedicated fiber optic lines. These systems are utilized by businesses and organizations with intensive Internet usage. In terms of cost, advanced DSL services are the most expensive. Originally, when many companies began entering the broadband market competition was viewed as a way to drive down costs, but company failures within the market have led to the consolidation of broadband companies. This consolidation reduced the number of available broadband providers, thus reducing competition within the high-speed telecommunications market.

Fixed wireless systems are another form of broadband telecommunications. These systems are delivered to customers through satellite feeds and towers. Wireless broadband infrastructure holds great promise for remote areas because of their potential endless reach, as well as reduced installation costs, as compared to landline systems. An example of an isolated town using such service is McDermitt, Nevada. McDermitt, Nevada is located on the Oregon/Nevada border with a population of 373. Unable to get a wireline company to supply the school system with service, they installed a two-way satellite service offered by Intellicom, offering 128k upload speeds and 2–4Mbps download speeds, at a cost of $1,700 a month (Goff 2001).

Given the different types of broadband telecommunications, the next step in the analysis is to examine broadband connectivity in a geographical context. From a global perspective, the deployment of broadband telecommunications clearly is driven by the demand for digital linkages between developed countries, with the United States as most connected. One of the key obstacles to understanding the spatial dynamics of broadband availability is the lack of comprehensive data for mapping broadband infrastructure due to competition among providers. But some states have managed to use political influence to encourage the participation of providers in mapping projects, such as Connect Kentucky (Connect Kentucky 2008).

Though the United States leads the world in broadband connectivity, numerous countries recognize its economic importance and continue to focus on the installation of broadband telecommunications. According to the

European Union (EU) Committee of the American Chamber of Commerce in Belgium, which represents the interest of over 140 American businesses, broadband telecommunications are a prerequisite for the economic development of Europe. The following excerpt from their official position paper details the stance of the EU Committee on the significance of broadband telecommunications:

> Broadband creates huge opportunities in terms of reduction of cost, efficiency, regional development, environmental, and social gains. The provision of broadband connectivity is no different from that of an effective road and rail network. The private and public investors in railways often did not foresee all of the benefits these networks offer today at the time when investments were made. Yet today, we couldn't live without them. Nowadays, multinational companies will not locate new facilities, or grow existing ones, in regions that do not have the necessary communications infrastructure, the same way that they will not do so in countries without the necessary transport infrastructure. Market players are responding to this by ensuring new business parks have state of the art communications links. Business will only deploy individual knowledge workers if they can engage in effective teleworking, for which broadband access to the home will be a prerequisite. Broadband is therefore a pre-requisite for economic growth and social inclusiveness across the EU (EU Committee 2002, 4).

In 2002, the national government of Ireland decided to invest 300 million (US$297 million) to bring broadband Internet technology to 123 towns. The three-phase program will bring 50,000 km of fiber networks to 19 towns by the end of 2003, 48 towns in 2004, and 56 towns in 2006 (Lillington 2002). The plan will create fiber-optic rings around towns with 90 percent paid by the national government and ten percent by local authorities. The ultimate goal of the program is to bring a five-megabit, always-on fiber optic connection to every home in the country by the end of 2005. Interestingly, the 90 percent national/ten percent local finance ratio utilized in the Irish broadband telecommunication program is the same formula used for the construction of the U.S. interstate highway system during the 1950s (Townsend 2002).

Despite the acknowledged importance of connectivity, developed and underdeveloped countries still lag behind the United States. In comparison with the U.S., Europe possesses a different landscape in regard to the distribution of broadband telecommunications. In Europe, for example, cable television has never approached the popularity achieved in the United States, thus an extensive cable network is lacking. The European system is also less competitive in terms of cable systems versus telecommunications companies for customers. In locations where broadband capable cable systems exist in Europe, they are often owned by the same telephone companies that have possession of the only other wire into the home (The McKinsey Quarterly

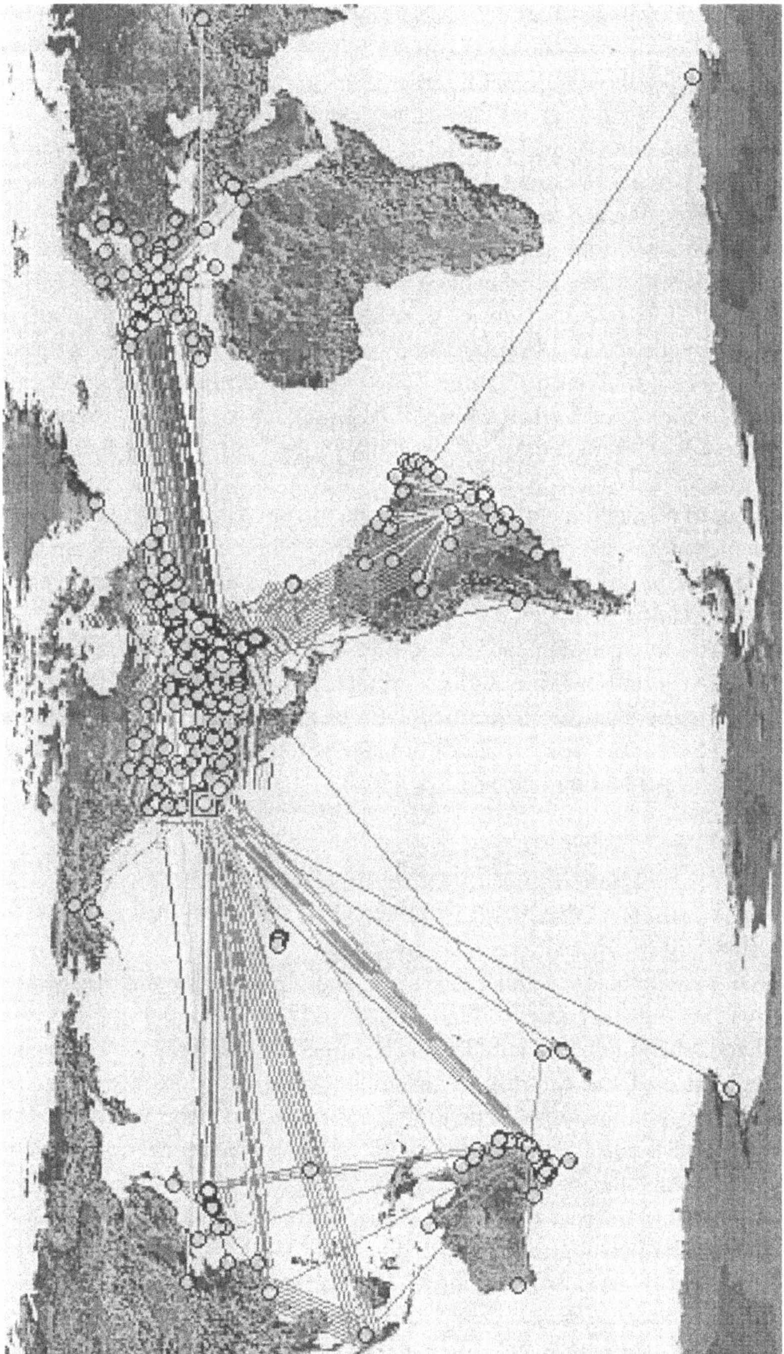

Figure 2-1. The Worldwide Commercial Internet. *Source:* Cooperative Association for Internet Data Analysis 2002.

2001). Underdeveloped regions, such as Africa, have fallen seriously behind in the number of Internet users. According to the CNN Virtual Villages Project (2005), although global Internet users grew to nearly a half billion in 2003, Africa only had 6.1 million users (less than 1 percent of the total online population). In comparison, the U.S. and Canada had 171 million users (roughly 34 percent of the total population).

Although recognized as the world leader in connectivity, broadband telecommunications are not available to every citizen in the United States. According to a report from the Federal Communications Commission (2008), areas with high population density have a strong positive correlation with the presence of high-speed subscribership, while areas of low population density have a strong negative correlation. Data from the study suggests stark differences in high-speed service based upon population density—though the report identifies 99 percent of all zip codes in the United States have access to some form of high-speed Internet. For example, in Arkansas approximately 30 percent of residential end users in the state do not have a single high-speed service provider, while New Jersey is covered with at least one form high-speed service provider. Table 2-1 provides state-by-state comparative data for the United States.

Broadband telecommunications exist in different forms and vary in geographic concentration. From a global perspective, it is apparent that broadband telecommunications is an instrument of developed countries. While in the United States, the availability of broadband telecommunications is highly dependent on population density.

THE INFORMATION ECONOMY AND BROADBAND TELECOMMUNICATIONS

Information technology, human resources, and globalization are the dynamic forces of the "new" economy (National Governors' Association 2000). The role of broadband telecommunications becomes increasingly important with the emergence of the Internet as the linking tool between businesses and people throughout the world. The global information economy benefits from what economists call "network effects," meaning the more technology that is deployed, the greater its value. As demonstrated in Figure 2-2, the continued global use of the Internet makes the service increasingly valuable for people and businesses choosing to connect.

Technological advancement is driving the structural transformation of the global economy (Berry, et al. 1997). The development of internationally integrated production and distribution systems, seen by many as the key

Table 2-1. Percentage of Zip Codes with High Speed Lines in Service as of June 30, 2007

	Number of Providers										
	Zero	One	Two	Three	Four	Five	Six	Seven	Eight	Nine	Ten or More
Alabama	0 %	0 %	2 %	6 %	10 %	15 %	15 %	13 %	10 %	10 %	18 %
Alaska	0	15	40	20	13	6	1	2	2	0	0
Arizona	0	0	0	1	5	12	10	9	10	8	46
Arkansas	0	0	3	9	20	24	18	8	6	6	5
California	0	0	1	5	10	10	7	7	8	8	44
Colorado	0	0	1	4	7	14	15	13	8	5	33
Connecticut	0	0	0	1	7	14	17	20	13	13	15
Delaware	0	0	0	3	10	10	26	19	10	9	12
District of Columbia	0	0	0	8	4	0	0	0	0	12	76
Florida	0	0	0	0	1	2	5	9	9	10	63
Georgia	0	0	1	3	8	13	15	13	7	7	33
Hawaii	0	0	2	10	10	17	13	17	19	10	2
Idaho	0	0	1	10	14	25	16	11	7	4	10
Illinois	0	0	2	9	13	17	15	10	8	7	19
Indiana	0	0	3	6	10	16	15	15	13	8	14
Iowa	0	3	10	14	16	17	12	10	7	3	8
Kansas	0	0	5	12	14	15	11	10	7	8	17
Kentucky	0	4	16	15	13	12	11	9	6	5	8
Louisiana	0	0	1	4	9	18	18	13	8	6	24
Maine	1	5	10	18	15	19	13	11	4	3	0
Maryland	0	0	2	4	12	18	10	9	6	4	34
Massachusetts	0	0	1	1	7	18	18	14	9	7	24
Michigan	0	0	0	3	6	11	15	15	13	9	28
Minnesota	0	0	3	12	16	15	10	10	8	5	21
Mississippi	0	0	3	6	14	20	15	9	8	7	18
Missouri	0	2	5	12	15	17	13	9	8	7	11
Montana	0	0	2	7	28	28	15	8	6	3	4
Nebraska	0	0	1	6	18	22	19	13	7	7	7
Nevada	0	0	0	3	10	16	10	13	5	9	34
New Hampshire	0	0	1	1	12	18	25	22	7	5	9
New Jersey	0	0	0	0	3	5	12	12	13	12	42
New Mexico	0	0	2	5	15	17	25	10	6	5	15
New York	0	0	2	6	13	14	15	12	9	7	20
North Carolina	0	0	0	3	6	10	12	15	11	11	32
North Dakota	0	4	26	31	24	8	3	1	1	0	1
Ohio	0	0	0	1	2	7	11	20	19	14	26
Oklahoma	0	1	4	8	17	17	13	11	10	8	11
Oregon	0	1	3	9	15	14	11	9	7	9	22
Pennsylvania	0	1	3	7	11	13	14	13	10	7	20
Puerto Rico	0	0	1	3	6	13	36	12	14	5	9
Rhode Island	0	0	1	7	7	15	12	15	14	14	16
South Carolina	0	0	0	2	6	9	12	13	12	12	33
South Dakota	0	4	23	24	19	12	8	4	2	2	2
Tennessee	0	1	4	6	9	14	11	11	9	8	27
Texas	0	0	1	2	5	8	12	14	13	10	35
Utah	0	0	3	2	9	16	15	13	7	2	33
Vermont	0	0	0	9	21	15	14	16	8	9	7
Virginia	0	0	1	5	11	15	14	14	10	7	22
Washington	0	0	1	4	12	12	11	10	8	6	36
West Virginia	2	11	18	20	17	12	9	4	3	3	1
Wisconsin	0	0	1	3	10	16	20	18	11	7	16
Wyoming	0	0	1	12	18	17	25	11	7	5	4
Nationwide	0 %	1 %	4 %	7 %	11 %	14 %	13 %	12 %	9 %	7 %	23 %

Source: Federal Communications Commission 2008.

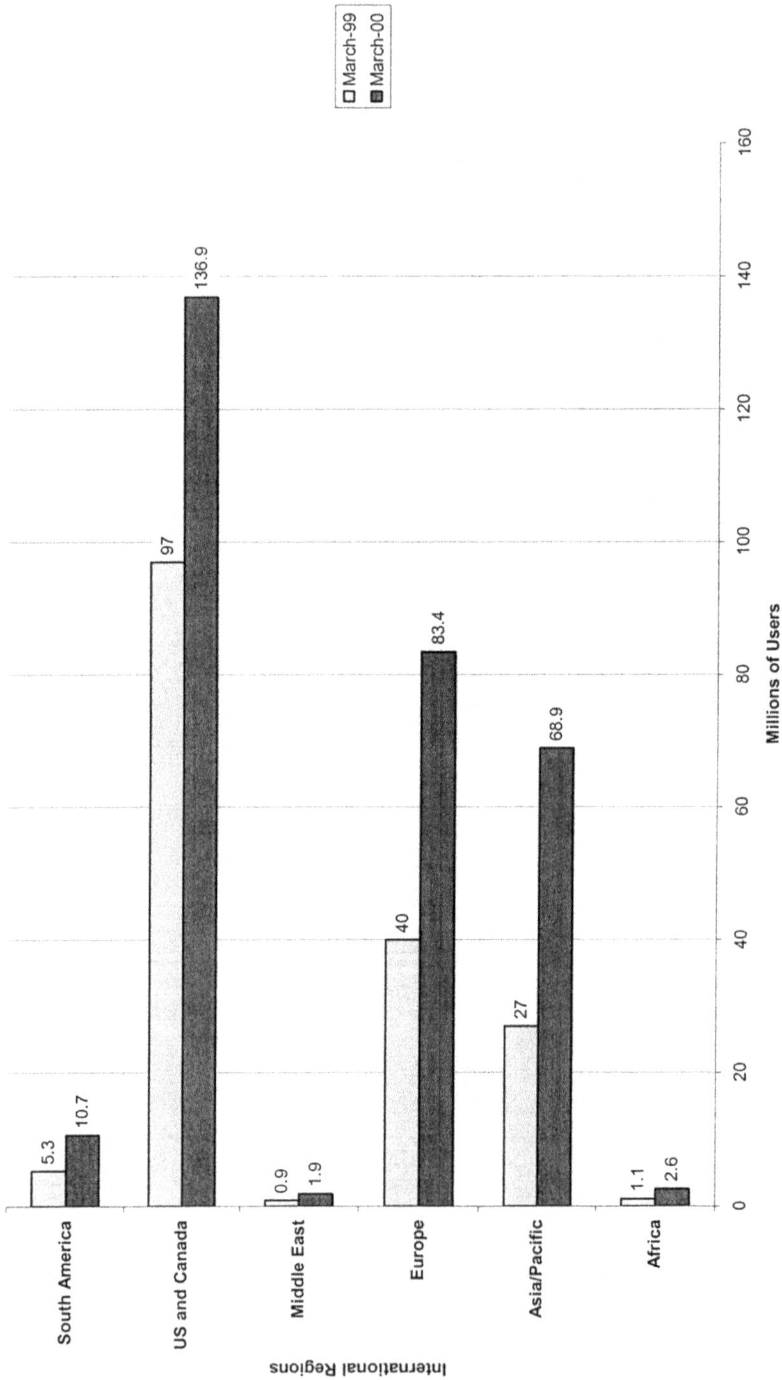

Figure 2-2. Internet Access Growth by International Region (in millions of users). ***Source:*** **NUA Internet Surveys 2002.**

feature of the global economy, has been a spatially uneven process due to the availability of information and communication technologies (Breathnach 2000). Just as important as the technological forces that change the structural nature of the global economy is the role of technology in increasing competitiveness among countries. According to Phillips (2000), the importance of information technologies cannot be overstated; it is the altering force behind the development process in the global economy. A key component of information technologies influencing economic development around the world is telecommunications.

The wide availability of relatively inexpensive high-speed telecommunications bandwidth is an important source of vibrancy in the global information economy. One of the most easily recognized global economic trends dependent upon telecommunications infrastructure is e-commerce. The evidence for the United States is that consumers who have high-speed "always on" Internet access spend four times as long online as dial-up users and engage in a far wider range of e-commerce activity (The Economist 2001). According to Grant (2002) global e-commerce will top $1 trillion in 2002, with business-to-business (B2B) transactions making up 83 percent of Web sales. The majority of B2B sales consist of large-volume purchasing, meaning companies are spending huge amounts of money with online transactions. So, as companies become more dependent upon B2B transactions for orders, the fate of businesses and potential suppliers who are not accessible to the B2B market will become increasingly noncompetitive.

Though information technology and telecommunications serve as the conduit between businesses and consumers, in the "new" economy knowledge is the key factor of production (Blodgett 2000). Within the information economy, competitive advantage centers on the ability to assemble the best human resources to utilize emerging technologies for business development. The nexus of competitive advantage is located in regions that can generate, retain, and attract the best talent (Florida 2000). The ability of a particular location to retain knowledgeable workers is commonly referred to as "quality of life" or "quality of place." These "quality" locations offer amenities, such as recreational and educational opportunities, which retain and attract skilled labor.

Austin, Texas, for example, is recognized as a city that offers high quality of life and boasts one of the highest concentrations of high-tech industries within the United States (Schantz-Feld 2000). By building on local institutions and locational advantages, such as the University of Texas-Austin and highly educated labor force, the Greater Austin Chamber of Commerce identified eight to ten clusters, including industries such as computers, software, semiconductors, film, music, telecommunications, multimedia, and bioscience (Smith and Milano 2000). In 2000, *Yahoo! Internet Life* ranked Austin as one of the top five "Most Wired Cities" in the United States. It is not unusual that locations rich in quality of life and high-tech clusters, such as Austin, Texas, have a high concentration of advanced telecommunications and Internet users. The Intelligent Community Forum of the World Teleport Association researches the role of broadband telecommunications in the economic development of "intelligent" communities, such as Austin, Texas and Research Triangle, North Carolina. These "intelligent communities" are defined as:

> Communities, whether within the political boundary of a municipality or comprising a larger region, that view bandwidth as the new essential utility, as vital to economic growth and public welfare as clean water and dependable electricity . . . the communities make conscious public-policy choices to seize the control of the community's broadband destiny in order to ensure their businesses and citizens access to advanced information and communication services (World Teleport Association 2002, 1).

Though there are many ways to examine the importance of broadband telecommunications in the information economy, such as location-based initiatives like Austin, Texas, perhaps the most imperative is to study its significance to business and industry. In 2000, *Area Development* presented a discussion forum of top CEOs of Fortune 500 companies on the impact of emerging technologies on the future of their industries. From Ford Motor Company to Hewlett-Packard, broadband telecommunications play an important role in their business plan and corporate structure. According to Chairman and Chief Executive Officer of Lucent Technologies Richard A. McGrinn, "Communications technologies have moved beyond mere business tools . . . they have become strategic assets and competitive weapons that make a difference (McGrinn 2000, 72)."

Carly Fiorina, President and CEO of Hewlett-Packard, believes that corporate profitability within the new economy will depend upon three fundamental principals: e-services, appliances, and infrastructure. E-services are the creation and delivery of Internet services to all computerized devices, far beyond the traditional capabilities of websites. Appliances are any devices that contain a microchip. These appliances become an "information appliance" when connected to the Web. The final principal is infrastructure. The proper infrastructure must be available to allow e-services and "information appliances" to function. According to Fiorina, it must be an infrastructure that "supports billions of devices and trillions of transactions . . . it has to be always on, always available, always reliable, and always secure" (Fiorina 2000, 67). As exemplified by the product development plans of Hewlett-Packard and others, broadband availability is necessary for using the new information-based devices.

Though significant to the "big business," the importance of broadband connectivity is not limited to Fortune 500 companies. According to a study by the Yankee Group (2001), a technology-consulting firm, an overwhelming number of small business owners consider broadband Internet access a significant, if not critical, resource. The study notes that 90 percent of the 550 small businesses that use SBC's digital subscriber line (DSL) service experienced gains from increased productivity meeting or exceeding the monthly cost of the service. The research also found businesses that used high-speed access for two years or longer were more likely to use the Internet for advanced capabilities such as distance learning, collaborating on electronic documents, or purchasing goods and services.

RURAL BROADBAND CONNECTIVITY AND ECONOMIC DEVELOPMENT: OPTIONS AND OPPORTUNITIES

Government leadership recognizes the importance of broadband telecommunications to the economic development of rural America. According to Representative Mike Pence of Indiana, Chairman of the U.S. House of Representatives Subcommittee on Regulatory Reform and Oversight, the role of broadband telecommunications in rural economic development is imperative. His comments from a congressional hearing entitled *Economic Development in Rural America, Small Business Access to Broadband* address the importance of broadband connectivity and identify rural areas as holding a competitive advantage in attracting businesses:

> Today communities that do not have broadband access to the Internet face the same barriers to economic development that communities, mostly rural, faced in previous generations when the mills, railroads, highways and airports passed them by. Without broadband access, rural communities will be unable to entice businesses that rely on the Internet to relocate and take advantage of the many qualities that rural communities offer. The other benefits, low crime, inexpensive housing, lack of traffic, clean air and a connection with one's neighbors are things that are missing in the booming metropolises of this country. All these things taken together are the competitive advantage of our small towns and of rural America at large (U.S. House of Representatives 2001, 6).

If rural communities are to foster economic development opportunities within the framework of the global information economy, they must possess sufficient telecommunications capabilities.

As noted earlier, not all rural areas of the United States have access to advanced telecommunications. Even though a number of rural areas of the United States are without broadband services, many communities are proactive in acquiring high-speed service through innovative strategies. Rather than waiting for a private provider to come to the rescue, local communities themselves are formulating their own answer for providing broadband network availability, with financial and technical assistance from such entities as state and local government and private sector industries. As noted by recent research (Bonnett 2000; Koo, Luger, and Stewart 1999), rural communities stand at different levels of readiness for broadband networks, while some are prepared for implementations, others are resistant to change and lack local leadership. Activities such as "needs assessments" and "demand aggregation" seem to be the most widely used techniques for preparing and encouraging local installation of broadband networks. Once a community determines the need for broadband network installation, it is important for the community to know all of the available providers of broadband networks. In general, telephone carriers are commonly recognized as the only broadband option, but there are several competitors in the market. In *Closing the Digital Divide in Rural America* (2000), Parker presents options for broadband network installation:

> *Telephone carriers.* If service is provided by a locally managed cooperative or small independent, then the chances of finding a satisfactory local solution are good, provided the local provider has access to construction funds from the Rural Utilities Service and is eligible for Universal Service Fund (USF) high cost support. If the local rural telephone provider is a large, urban-based telephone company, they might not be responsive, even if the rural community could demonstrate enough demand to show that service would be profitable. The carrier may believe they can get a greater return on their investment elsewhere.
>
> *Cable Television.* If the local telephone carrier is not responsive, then the local cable television company may be a viable alternative. They may be able to upgrade their network to provide local broadband digital service through cable modems.
>
> *Electric Utilities.* A particularly promising source for potential rural broadband solutions is the local electric utility. Power companies have internal needs for data communications within their power network. They have rights of way and access to all businesses and residences. Fiber optic communication works better along power lines than traditional copper telephone lines. The power lines interfere with the electrons in the copper wire but leave the photons in the fiber optic channel unaffected. Once fiber is installed in the power grid, the excess capacity that fiber brings could easily accommodate most other rural needs in their service areas.

> *Competitive Providers.* It is less likely in rural than in urban locations, but in some instances a competitive provider may be found to offer service. It may be a local Internet Service Provider (ISP) frustrated by the lack of local broadband capability that upgrades to CLEC (Competitive Local Exchange Carrier) status in order to offer services that would not otherwise be available.
>
> *Local Government.* Either pre-existing or newly formed governmental entities may be a potential solution in some rural communities. There is a longstanding rural tradition of local people organizing themselves as a last resort method to obtain services not otherwise available. One good example is the western tradition of rural cooperatives bringing power to rural communities that no private sector power company wished to serve. Rural telephone cooperatives have become the telephone service provider in many rural communities for the same reason: no for-profit company was willing to make the investment.

As noted by Parker, many options exist for communities to gain access to advanced telecommunications. But, once available, how can communities turn broadband into economic development?

Although many strategies are used to encourage economic development in rural communities, one of the most noted in recent years is the recruitment of telecommunications-based service firms. This style of rural economic development, often referred to as "teleconomic development," involves the recruitment of back-office operations, such as customer-service centers and telemarketing, to rural communities to take advantage of available, low-cost labor (Nelton 1994). For example, in the CERAtown of Powell, Wyoming, population 5,300, LJS Global Information Services provides back-office support to such companies as Reuters and Dow Jones. The company processes over 70,000 corporate and U.S. treasury debt securities work in excess of $6 trillion (Maturi 2001). According to Matt Scheicher, Chairman and CEO of LJS Global Information Services, "Powell is an excellent place to conduct high-tech information services . . . we are wired to the world . . . and the quality work force and lower operating costs in Wyoming allow us to compete in both price and service" (Maturi 2001, 81). Although Powell, Wyoming is just one case, the availability of low-cost, high quality labor is the key ingredient for determining location and profitability. Yet, the linking mechanism that enables this labor to be tapped for employment is advanced telecommunications capability.

Another highly recognized rural economic development strategy is the initiative of telecommuting or teleworking. Telework, often referred to as telecommuting, is defined as the situation where paid workers reduce their commute by carrying out all or part of their work away from their normal places of business, usually from home (Innovisions Canada 2001).

According to Odasz (2001), studies suggest that rural telecommuters have fewer turnovers, less absenteeism, and a stronger work ethic than workers in urban areas. This type of quality employee creates a real incentive for businesses to consider employing rural citizens. Based on numerous case studies, some jobs are better suited to telecommuting than others, including such positions as medical transcriptionists and insurance underwriters (Valley Metro 2002). Yet the possibilities of creating employment in rural areas through telecommuting are dependent upon advanced telecommunications. According to Quay (1995), telecommuting strategies hold great potential, but only with sufficient telecommunications infrastructure:

> Telecommuting has the potential to alter this equation. With telecommuting, it matters little whether you are 30 miles or 300 miles away from the office. Once the locational decision of home is separated from the physical distance of work, the options for where home is become enormous. One possible future this could lead to is the re-ruralization of America. Telecommuting (for rural areas) will require enhanced communications infrastructure in rural areas (Quay 1995, 3).

Both "teleconomic development" and telecommuting strategies are in need of "enhanced" telecommunications before they can be seen as possibilities for rural economic development. These ideas represent only a limited perspective on potential applications of broadband telecommunications for rural job creation. According to Tassey (2002), the emergence of e-commerce will eventually reach the smallest of manufacturers within the knowledge-based economy. The importance of telecommunications for information processing will not only serve as important to rural job creation strategies, but for existing business to survive within the ever changing context of the global economy. Rural areas are not exempt from this movement toward information-based industries, businesses, and products.

CONCLUSION

Based on the literature, the potential of economic growth within the context of a global, information-based economy is dependent upon connectivity through advanced telecommunications. According to U.S. Federal Communications Commissioner, Michael Copps, "The transformative potential of broadband technologies is, I believe, akin to the major infrastructure developments that built America to greatness (Macklin 2002, 1). As pointed out by Thomas Friedman in *The Lexus and the Olive Tree*, "jobs, knowledge use, and economic growth will gravitate to those societies that are the most connected, with the most networks and the broadest amount of bandwidth" (Friedman 2000, 73).

Broadband telecommunications are a critical component of the "new" economy. For all of the intricate components of the information economy, from business to business e-commerce to available labor in Powell, Wyoming, there must be a mechanism of connectivity to allow them to operate and become involved. That mechanism is broadband telecommunications. Thus, just as the interstate highway played an important role for movement of goods within the "old" economy, broadband telecommunications serve as the transportation system of the "new," information-based economy.

Chapter Three

Federal and State Broadband Policies: Shaping the Prospect of Economic Development in Rural America

The importance of government intervention is recognized in major theoretical contributions relating to economic development, including Perroux (1950), Friedmann (1966), Myrdal (1957), and Hirschman (1958). Porter (1990) notes that one of the most important and traditional roles of government is to create and upgrade factor conditions, such as skilled human resources, basic scientific knowledge, economic information, and infrastructure. The economic development literature supports the idea of government intervention in assuring access to physical infrastructure systems, including telecommunications (Porter 1990; Romer 1990; Caristi 2000). In the United States, the Telecommunications Act of 1934 serves as the original legislation for assuring telephone access to all citizens, with particular attention to rural people.

As exemplified in Chapter Two, the participation of government is important in assuring the timely deployment of broadband telecommunications to rural Americans. If broadband telecommunications is a key ingredient for the economic development of rural areas in the information economy, government policy must ensure that universal access to advanced telecommunications is guaranteed.

The rural broadband deployment policy of the United States government follows the belief that increased competition encourages service into underserved areas, which is exemplified with the passage of the Telecommunications Act of 1996. This legislation supports additional competition among private suppliers, far beyond the original Telecommunications Act of 1934. Recommendations from the U.S. Department of Agriculture study, *Advanced Telecommunications in Rural America: The Challenge of Bringing Broadband Service to All Americans* (2000), suggests privatization as the correct path, but other policy ideas must be formulated to assist rural areas in gaining access to broadband telecommunications. The study states:

> Policymakers should promote competition, where possible. Using the pro-competitive provisions of the Telecommunications Act, some competitive local exchange carriers have deployed advanced services in rural areas of the Country. Some wireless carriers have also indicated an interest in providing voice and high rate data, especially if universal service policies can be reformed. Competition leads to lower prices, more customer choice, rapid technological advances, and faster deployment of new services. Given unique challenges faced by rural Americans, however, other government policies must be considered as well (USDA 2000, 8).

The policy of competition is working for urban Americans, but it is not meeting the needs and demands of rural citizens (US Dept. of Commerce 1998; Drabenstott and Schaeff 2001). Unless additional policies are put in place to compensate for the failures of the pro-competition policy for rural areas, they will continue to be left behind. Based on the success of public-private partnerships for broadband telecommunications, cooperation, not competition, seems to be the most logical policy solution for rural broadband deployment (Youtie 2000).

The purpose of this chapter is to better understand the role of federal and state policy in the deployment of broadband telecommunications in the United States. If broadband telecommunications is to be examined as a possible tool for encouraging economic development in rural communities, it is imperative to understand the policies that support or deter its deployment. This section provides an important foundation for the remainder of the research because policy preparation is vital in the development of a model for rural economic development.

If federal and state government involvement is essential to ensure broadband telecommunications access in rural and underserved areas, then the following questions are important in the overall context of this research:

1) What are the identifiable policy trends at both the federal and state levels with regard to broadband telecommunications in rural areas?
2) Can innovative policies be incorporated into a model for rural economic development based upon broadband telecommunications?

The data used to address these questions originate from the Congressional Record of the 106th and 107th Congress and a fifty-state survey of legislative and executive proposals.

BROADBAND INTERNET ACCESS: FEDERAL POLICY INITIATIVES

Broadband telecommunications infrastructure is predominantly deployed by the private sector throughout the United States. According to a survey of the Federal Communications Commission (FCC), as of June 30, 2000, there were 2.8 million broadband lines in the United States and roughly 10.7 percent of online households (about 4.5 percent of all U.S. households) had broadband access (FCC 2000). The future of the broadband market is full of uncertainty, as competing companies and industries try to anticipate technological advances, consumer preferences, and even cultural and societal trends.

From a national policy perspective, the goal is to ensure that broadband deployment is available to all sectors and geographical locations of American society (Kruger and Gilroy 2000). The Telecommunications Act of 1996 specifically addresses the deployment of broadband telecommunications in Section 706 of the Act. It requires the FCC to determine whether advanced telecommunications capability is being deployed to all Americans in a reasonable and timely fashion. If this is not taking place, the Act requires the FCC to take immediate action to accelerate deployment of such capability by removing barriers to infrastructure investment and by promoting competition in the telecommunications market.

In response to deployment concern expressed in Section 706 of the Telecommunications Act of 1996, the FCC adopted Report 99-5 in 1999 pursuant to Section 706. The report states: "the consumer broadband market is in the early stages of development, and that, while it is too early to reach definite conclusions, aggregate data suggests that broadband is being deployed in a reasonable and timely fashion." In 2000, the second FCC 706 Report, 00-290, continued to support the idea that broadband services were being deployed in a timely fashion, but noted that certain groups were vulnerable to not receiving broadband service in a timely fashion. Those identified as "vulnerable" included rural, minority, low-income, tribal areas, and inner city consumers.

While the FCC's position is not to intervene at this time, some assert that legislation is necessary to ensure fair competition and timely broadband deployment. Currently the debate centers on two specific concerns: 1) compelling cable companies to provide "open access" to competing Internet service providers, and 2) easing certain legal restrictions and requirements, imposed by the Telecommunications Act of 1996, on incumbent telephone companies that provide broadband access (Kruger and Gilroy 2000). Although the Telecommunications Act of 1996 serves as the major regulatory framework for the telecommunications industry, the Congress offered numerous pieces of legislation in response to concerns of "open access" and easing legal restrictions for incumbent telephone companies with regard to broadband telecommunications. The recent federal policy initiatives can be divided into two distinct themes: 1) pro-competition policies and 2) rural specific policies.

The majority of federal policy addressing broadband telecommunications deployment is aimed at assuring fair competition among broadband suppliers. The most obvious policy concern is the openness of the broadband market to alternative providers, such as cable television companies, which is exemplified in House Concurrent Resolution 173. The concurrent resolution expresses the Congressional opinion that the Federal Communications Commission should exercise its authority under the Communications Act of 1934 to ensure that unaffiliated service providers have open, nondiscriminatory access to broadband facilities that enable access to the Internet over cable systems. This feeling is carried out through several pieces of legislative action, including the "Internet Growth and Development Act of 1999," the "Internet Freedom Act of 1999," the "Internet Freedom and Broadband Deployment Act of 1999," and the "Consumer and Community Choice Access Act of 1999."

The "Internet Growth and Development Act of 1999," on record as House Resolution 1685, prohibits anticompetitive or discriminatory behavior by broadband access transport providers that have market power and would deem as a violation of antitrust law any preferential treatment given to affiliated Internet Service Providers (ISPs). The resolution amends the Communications Act of 1934 to lift restrictions on data transmission by incumbent telephone companies and to remove, under certain conditions, unbundling and resale requirements. A nearly identical piece of legislation, House Resolution 1686, commonly referred to as the "Internet Freedom Act of 1999" prohibits anticompetitive or discriminatory behavior by broadband access transport providers that have market power and would deem as a violation of antitrust law any preferential treatment given to affiliated ISPs. To summarize the major contributions of these two pieces of legislation, they specifically prohibit broadband providers from offering preferential treatment to certain ISPs, thus assuring fair pricing and access.

In addition to the assurance of open access and non-discriminatory pricing, federal policy also addresses the regulatory stance of the federal and state governments with regard to broadband access. House Resolution 2420, known as the "Internet Freedom and Broadband Deployment Act of 1999" amends the Communications Act of 1934 to prohibit any states or the FCC from regulating the provision of high-speed data services. The legislation lifts restrictions on data transmission by incumbent telephone companies while also removing unbundling and resale requirements. The act also requires incumbent local exchange companies to provide any ISP with the right to interconnect with the carrier's high-speed data service.

The final step to assure a competitive marketplace for the deployment of broadband service is the regulation of cable operators by the FCC. House Resolution 2637, the "Consumer and Community Choice in Access Act of 1999," amends the Communications Act of 1934 to give the FCC authority to require cable operators to provide functionally and economically equivalent interconnection to all ISPs, regardless of their affiliation with the cable operator. Until open access or interconnection standards are established, it would enable ISPs to obtain access through leased commercial access provisions contained in Section 612 of the 1934 Communications Act.

The second significant trend in federal policy is the offering of assistance to encourage broadband access to rural Americans in a timely manner. As noted in the introduction, the FCC feels that broadband is being offered in a "timely manner," but Congressional legislation suggests that broadband access is not moving as quickly as they would like. Based upon the survey of legislation, the policies addressing broadband deployment to rural America fall into two discernible themes: 1) grant and loan programs and 2) tax credits for private broadband suppliers.

Numerous pieces of legislation offer grant and loan programs to encourage rural broadband deployment. Both House Resolution 4122 and Senate Bill 2307, known collectively as the "Rural Broadband Enhancement Act of 2000," encourage broadband deployment in rural America through the FCC's Universal Service Fund and the U.S. Department of Agriculture's Rural Utilities Service (RUS) Loan Program. In addition to the traditional loan programs, grant funding is also offered to assist rural areas. House Resolution 4477, the "Digital Bridge Trust Fund Act of 2000," establishes a fund for programs to improve skills and career opportunities in information technology for underserved communities. This legislation also includes funding for assisting in the deployment of local broadband networks in rural and underserved areas. Senate Bill 2097, the "Launching Our Communities' Access to Local Television Act of 2000," authorizes loan guarantees in order to facilitate access to local television broadcast signals in unserved and underserved areas. The bill also designates "high-speed Internet access" as eligible for loan guarantees.

Though several loan and grant programs are proposed, the most comprehensive grant and loan legislation for rural broadband access is House Resolution 5069, entitled the "Comprehensive Rural Telecommunications Act of 2000." It provides tax credits, universal service support, and RUS loans for the construction of broadband telecommunications facilities in rural areas. The legislation also includes incentive provisions for the encouragement of rural teleworking. Also introduced in 2001, were Senate Bill 966 and House Resolution 2038. This legislation gives the RUS, in conjunction with the National Telecommunications and Information Administration (NTIA), new authority to make low interest loans to companies that deploy broadband technology to rural and underserved areas. The loan programs provide for a "technology neutral" application process, meaning companies that want to serve rural areas may do so using satellite, telephone, cable, wireless or any other technology that gets the job done for a given area (Newsbytes 2001).

As a supplement to grants and loans, tax credits for private broadband suppliers make up a noteworthy amount of federal policy. House Resolution 4728 and Senate Bill 2698, known as the "Broadband Internet Access Act of 2000," provides a 10 or 20 percent tax credit for five years to companies investing in broadband facilities or equipment to serve rural and low-income areas. Senate Bill 2321, the "Rural Telecommunications Modernization Act," provides a 10 or 15 percent tax credit for three years to companies that invest in broadband facilities in rural areas. Based on the specifics of the legislation, the tax rate is dependent upon the type of equipment as well as the transmission speed of the system. As exemplified in the "Community Renewal and new Markets Act of 2000," the legislation calls for a broadband Internet access tax credit for five years to companies investing in broadband facilities or equipment to serve rural and low-income areas. The bill offers a 10 percent tax credit for "current generation" broadband service (defined as download speeds of at least 1.5 million bits per second), and a 20 percent tax credit for "next generation" broadband service (defined as download speeds of at least 22 million bits per second).

Based upon the review of federal broadband policies, their immediate concerns are fair markets and timely access to all Americans. Although federal policies are a critical component of both broadband regulation and deployment, it is important to understand the role of state policy in escalating opportunities for rural access to high-speed telecommunications. Unlike the federal government, states are less concerned about fairness and equitability across borders. Thus, based upon the review of proposed legislation, state policies are more creative in finding solutions to overcoming broadband accessibility problems in rural communities.

RURAL BROADBAND POLICY: A SURVEY OF STATE LEGISLATION

In order to better understand the use of public policy for broadband deployment to rural areas, this section examines legislation and executive branch initiatives for each of the 50 states from 1998 to 2005. It is the purpose of this extensive survey to identify trends in state broadband telecommunications policy, with particular focus upon legislation concerning rural areas. This

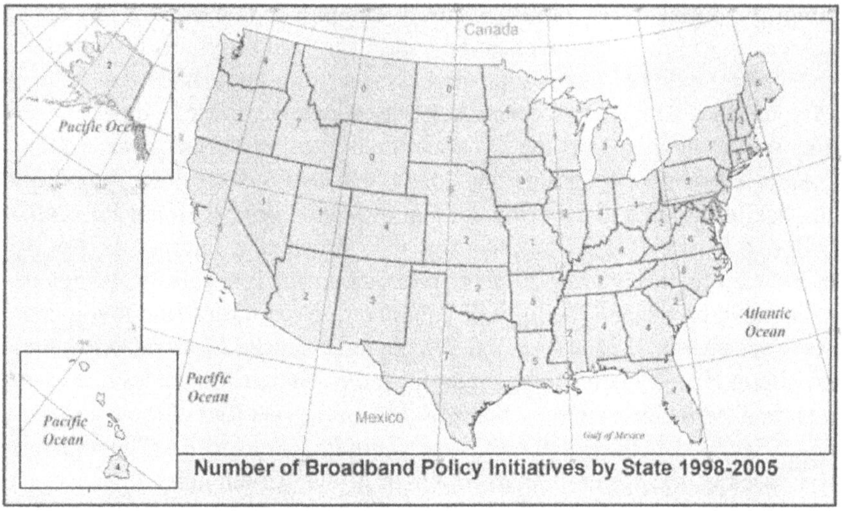

Figure 3-1. Number of Broadband Policy Initiatives by State 1998–2005.

section seeks to address the research questions through the following steps: 1) identify the number of rural broadband policies by state; 2) recognize trends in policy development; and 3) develop a summary of the legislative approaches addressing the broadband accessibility in rural areas for use in the development of a rural economic development planning model.

The number of policy initiatives at the state level is a convenient way to measure the awareness of political leaders to the importance of broadband telecommunications for rural citizens. Of the 50 states surveyed from 1998 to 2005, 47 were actively involved in rural broadband policy development. The leading states in policy formulation were North Carolina (8), Idaho (7), Texas (7), California (6), and Virginia (6). Figure 3-1 displays the activity of the entire 50 states from 1998 to 2005.

Beyond the mere number of policies, this research surveys the content of state legislation for creative policies to integrate into a model implementing broadband telecommunications for rural economic development. Furthermore, it is important to overview each piece of legislation not only to identify the activities of states, but also to identify broadband policies themes which complement the activities of local initiatives. These local initiatives, as discussed in Chapter Five, include leadership, institutions, and economic development planning.

Based upon the 50-state survey of broadband specific legislative and executive programs, six policy themes emerge. These broadband related policy themes are grouped into the following categories:

1) Administration, Assessment, and Planning
2) Education-focused Initiatives
3) Tax Credit Initiatives
4) Loan and Grant Initiatives
5) Broadband Industry Related Legislation
6) Economic Development Initiatives

Administration, Assessment, and Planning

The administration of information technology (IT) is of importance to numerous states. One important administrative policy trend is the creation of IT-specific offices within executive branches of government to oversee issues related to information technology. This trend is exemplified through legislation from the states of Alabama, Arkansas, and North Carolina.

In Alabama, Senate Bill 358 (SB358), introduced on February 17, 2000, created the Office of Information Technology as a department of the executive branch of government to provide for the administration of programs and services related to the operation and implementation of IT services in Alabama. Based upon this legislation, the administrative focus of the office is to ensure that the children of Alabama citizens have access to technology in the public schools and public institutions of higher learning. In addition to the creation of an executive department, SB358 created the Information Technology Council to assist the Office of Information Technology, primarily through assisting in the development and implementation of the statewide information technology plan. The council consists of various members of the educational community, including seven representatives from four-year public universities appointed by the Council of College and University Presidents, three classroom teachers, two two-year college presidents, three representatives from the State Department of Education, four at large members to be appointed by the Speaker of the House, the President of the Senate, and President Pro Tempore of the Senate. In addition, the Executive Director of the Alabama Commission on Higher Education and the Chancellor of the two-year college system serve as ex-officio members.

In a similar 2001 executive proposal, Governor Mike Huckabee of Arkansas sought legislative support to establish a Chief Information Officer position. The focus of the position is to garner the necessary resources to build a strong technology infrastructure that will provide high-speed connectivity to every Arkansan (James 2001). According to Huckabee, the backbone facilitates the recruitment of high-tech businesses to Arkansas.

In 1999, the North Carolina Legislature introduced House Bill 1578 which repealed the powers and duties of the Secretary of Commerce relating to "information technology matters" and reestablished the duties within the Office of the Governor. The "information technology matters," which includes broadband telecommunications, are the duties of a newly created State Chief Information Officer (CIO) position. The CIO is appointed by the Governor and approved by the Senate.

In addition to the administration of information technology issues, assessments of broadband telecommunications infrastructure and usage appear in various states. Several legislative and executive proposals seek to identify the physical location of broadband telecommunications infrastructure, which includes both wireline and wireless applications. In addition to physical location, assessments also deal with issues of disparity in broadband service along socio-economic lines, as well as actual broadband applications, such as e-commerce.

Georgia Senate Resolution 626, introduced February 25, 2000, created the "Joint Task Force on Technology Disparities in Telecommunications" to identify and investigate the existence of disparities in the distribution and availability of telecommunications technologies and services across community, geographic, political, and regional lines. The task force examines the distribution and availability of state-of-the-art communications systems and services, such as cellular telephone, Internet access, and cable television. The task force also addresses the existence or nonexistence of competition within the telecommunication industries. Likewise, the Kentucky legislation titled Concurrent Resolution 327 directed a study of unequal availability of high-speed Internet access throughout the Commonwealth of Kentucky. The legislature noted that deployment of a high-speed backbone is not taking place in many rural areas. The study identifies areas of the state where DSL and ISDN are currently unavailable, determines the reasons for the slow diffusion, and recommends ways to improve offerings to rural areas. In 2004, Kentucky House Bill 627 established the Kentucky Broadband Task Force to examine regulation, cost, access to facilities, and competition. The task force consists of representatives from state government, local carriers, local Internet Service Providers, Municipal Utilities, etc.

Louisiana is another state planning for the future of telecommunications; one way is through the Louisiana Broadband Advisory Council established by House Bill 1265 in 2004. The council is directed by the Governor's Office of Rural Development and its purpose is to assist with economic development, particularly in rural areas. Universal connectivity is cited as a need in order for the state to attract new businesses and expand current businesses along with providing a better education for its residents.

In 2000, Maine Legislative Resolution 2039 called for the Maine Science and Technology Foundation (MSTF) to determine the physical condition of the information technology infrastructure in the state. The study directs the MSFT to:

1) Draft a request for information on IT infrastructure and disseminate the request to telecommunications service providers, Internet service providers, wireless service providers, and digital satellite and cable television companies.
2) Identify specific geographic areas where advanced telecommunications capability is deployed at a significantly lower rate than the rest of the state.
3) Study the investment in telecommunications facilities with advanced capabilities in rural areas, including an assessment of various levels of capability deployed under different technologies, the bandwidth capabilities of the deployment, and whether or not comparable bandwidth is deployed consistently.

In a similar piece of 2000 legislation, Pennsylvania House Resolution 130 directed the Joint State Government Commission to study the broadband and high-speed telecommunications infrastructure, both wireline and wireless, throughout the state. To address the issue quickly, the committee must report their findings, recommendations, and any proposed legislation to the General Assembly within six months of the adoption of the resolution.

Beyond the "what and where" aspects of broadband assessments is the concern of state governments as to the impacts of broadband usage, such as e-commerce. In 2001, Idaho House Concurrent Resolution 31 directed the appointment of a committee to study electronic commerce, broadband connectivity, and other related technologies. The committee reviews current policies and proposes legislation to ensure that electronic technologies continue to advance in the socio-economic environment, benefiting citizens, businesses, communities, and the state of Idaho. Likewise, Louisiana House Study Request 24 of 1999 requested a study to better understand the impact of Internet commerce and broadband telecommunications on the economy of the state of Louisiana.

Though assessments address various issues of importance, they appear to have significant value in broadband telecommunications deployment. In North Carolina, through an agreement between the state and its three largest telecommunications companies, BellSouth, Sprint, and GTE, every citizen of North Carolina is guaranteed broadband Internet access by 2003. According to Tom Runkle, Chief Planning Officer of North Carolina Office of Information Technology Services, having an inventory of the State's telecommunications infrastructure proved critical during negotiations. Runkle stated,

"Knowing what [infrastructure] you've got, in detail, is very important. You can't get anywhere without knowing that. It's grunt work. Let me tell you, it's hard to get that inventory" (Peterson 2000, 1).

The third theme addressed is the identification of policies related to broadband telecommunications planning for rural areas. In general, the broadband planning legislation focuses on preparing future legislation relating to rural telecommunications and measuring the success of previous planning efforts.

In Idaho, the legislature used planning studies to prepare for the complicated issues related to the regulation of the telecommunications industry. House Concurrent Resolution 7 appointed a committee to study the subject of the regulation of the telecommunications industry. The legislation references the purpose of the Telecommunications Act of 1996 and notes that the diversity of urban and rural population within the state makes the regulation of the telecommunications industry one of the most important public policy issues facing Idaho in the 21st Century.

In 1999, the Utah legislature, with the passage of Senate Bill 188, established the Rural Telecommunications Task Force within the governor's office. The task force is composed of 11 members and charged with reviewing and making planning recommendations on the following issues:

1) Identify eligibility criteria for Universal Service Fund Support for capital investment in broadband data services in rural areas.
2) Identify areas of the state where state government should assume liability for the costs of relocating facilities in the case of right of way realignments in order to encourage deployment of digital infrastructure.
3) Develop other possible solutions to aid in the deployment of advanced telecommunications in rural areas.

Tennessee Senate Joint Resolution 134 directed the Department of Economic and Community Development to conduct a planning study addressing the provision of adequate wireless telecommunications services in rural and other low-population areas throughout Tennessee. The following issues were explored:

1) The costs of erecting additional towers to enable wireless telecommunications service to be more accessible in rural areas
2) The feasibility of state and local governments participating in the erection of additional towers in the interest of improving wireless telecommunications for police, fire, and emergency personnel

3) The feasibility of participating in federal initiatives, interstate compacts, and public-private partnerships that would facilitate the erection of new towers in rural areas
4) All other aspects of wireless telecommunications service that might affect rural areas

Although Idaho, Utah, and Tennessee only recently authored telecommunications-related planning studies, other states are appraising the success of planning efforts. In Texas, a piece of 2001 legislation entitled the "Texas Universal Broadband Act of 2001" places success measures on the progress of broadband telecommunications throughout the state. It establishes a precise set of criteria for measuring the success of state efforts to deploy advanced telecommunications services. The act also introduces a specific component that allows for the use of state telecommunication funds for rural deployment and increases state funding for deployment of advanced telecommunications services.

Colorado proposed a bill in 2004 that would create the "Telecommunications Authority Law." This law would allow local governments to work together to create telecommunications authorities in areas that are underserved or unserved altogether.

Education-focused Initiatives

Based on the survey of legislation, broadband strategies incorporating education institutions, such as public schools and universities, are routine. Though educational institutions are involved, state legislation varies from attaining broadband access in schools to fostering public-private partnerships that improve broadband access. Just as the legislation varies, the broadband needs of states differ as well.

During the State of the State Address on February 6, 2001, Alabama Governor Don Siegleman noted the need to expand broadband Internet technology to public schools and offered a plan to connect every school to the Internet. In order to ensure that every Alabama student has access to the resources needed for a 21st Century education, Governor Siegleman committed to the goal of ensuring that every student in the state of Alabama will have high-speed broadband access to the Internet in their school by 2003. Currently, more than 50 percent of Alabama's schools have only dial-up Internet services or no connectivity at all (Office of the Governor of Alabama 2001). In contrast to Alabama, Arkansas Senate Bill 843 requires public schools in underserved rural and poor areas of Arkansas to utilize free access to broadband services.

The bill notes that although free broadband Internet is available in some areas, it is apparent that not all public schools are utilizing the technology. The bill requires schools to design programs that implement the access within six months of legislation passage.

Another governmental proposal offers tax incentives and discounts to broadband suppliers who provide broadband service to state agencies, which includes educational institutions. The Texas legislation, House Bill 2128, originally introduced in 1995, established the Telecommunications Infrastructure Fund (TIF) along with incentive regulation. Participating companies were required to file special discounted rates for connecting certain public entities, and those rates could not be increased for six years. The bill also included provisions for serving rural areas and educational institutions with high percentages of economically disadvantaged students. A follow-up piece of legislation in 1999, Senate Bill 560, eliminated several provisions established in House Bill 2128 that would have expired on the sixth anniversary of a company's participation date. However, the legislation did not extend the incentive regulation rates and the initial discount rates that expired in 2001. The bill allowed incumbent local exchange carriers who did not participate in the incentive program to receive reimbursement through the universal service fund for reduced rates for K–12 public schools and other public entities.

Legislation is not only seeking to address the needs of educational institutions themselves, but also uses educational institutions as key players in the dispersion of broadband access to communities and citizens. For example, in Florida, through a 2000 legislative initiative, $500,000 in state funding established the Community High-Technology Investment Partnership (CHIP) at Florida A&M University. The purpose of the legislation is to provide access to computers and Internet services for underserved youth in distressed communities, including urban and rural areas. Governor Jeb Bush directed these state resources to be used to expand into the PowerUP Florida Initiative. PowerUP is a collaborative and inclusive effort that gains its power from its partners: corporations, non-profits, and government organizations which ensure that America's underserved youth acquire the skills, experiences, and resources they need to succeed in the digital age. In partnership with the national PowerUP organization based in Virginia, the Florida-based group will establish at least 25 computer lab sites by the end of 2001 in underserved communities across Florida. The PowerUP Florida partnership includes Florida A&M University's Institute on Urban Policy and Commerce, the Governor's Office, the Department of State, IT (Information Technology) Florida, National PowerUP, private industry leaders, and the Searcy Foundation (State of Florida 2000).

Another university-based initiative is the Center to Bridge the Digital Divide at Washington State University. Established in 2001 by Governor Gary Lock and U.S. Senator Patty Murray, the main concern of the center is to speed deployment and use of digital technologies in underserved communities. Pilot test projects will provide high-speed Internet access to three rural communities north of Spokane. The program embodies the concept of using university outreach and research capabilities to benefit the development of rural communities.

Tax Credit Initiatives

Following in the path of federal policy, numerous states adopted similar plans for encouraging broadband deployment to rural areas through tax credit incentives for cable and telecommunications companies. The most common form of tax incentive is a credit toward broadband equipment. For example, Hawaii House Bill 275 provides an income tax credit for investment in broadband equipment. The bill offers a tax credit of three percent on qualified purchases for broadband equipment. Also included in the credit are wireless applications of broadband service, which may receive credits toward transmitting antennas, etc.

Although the use of tax credits for broadband equipment is a generally accepted procedure, differences exist in the amount of credit offered. This differentiation is generally dependent upon the type of equipment installed. The states of North Carolina and Virginia have authored nearly identical bills to encourage broadband investment through tax inducements.

In 2001, the North Carolina House and Senate passed the "Broadband Internet Access Tax Credit." The bill provides a credit to businesses for broadband equipment expenditures that expand high-speed telecommunications to underserved or rural areas. For current generation broadband equipment, a 10 percent credit can be taken on "qualified expenditures incurred with respect to qualified equipment offering current generation broadband services to rural or underserved subscribers." For next generation broadband equipment, a 20 percent credit can be taken. The maximum amount of tax credits allowed is $750,000 per tax year.

Similar legislation from Virginia, entitled the "Broadband Internet Tax Credit," provides a tax credit to applicable businesses in order to accelerate the deployment of current generation broadband to low-income and rural areas and next generation broadband to all citizens of Virginia. The tax credit is 10 percent on qualified equipment to supply current generation broadband services to rural and low-income areas. The tax credit is 20 percent on qualified equipment to deploy next generation broadband anywhere in

Virginia. The maximum amount of the credits allowed under the legislation is $750,000 per tax year.

Arkansas proposed similar legislation in 2005 with the "Broadband Incentive Act of 2005." This legislation provides a tax credit of 15 percent of the cost incurred to provide rural and underserved areas broadband access to broadband providers.

In addition to federal-style tax credits, some states authored legislation offering credits for the purchase of computers and Internet subscriptions. In the State of Arkansas, House Bill 2466 of 2001 offers a computer sales tax holiday for the purchase of personal computers and software. It is the logic of the tax holiday that low computer ownership of Arkansas is detrimental to the future prosperity of the state, and the digital divide will continue to grow unless citizens of Arkansas increase computer usage and technology skills development. The tax holiday occurred on August 4, 2001 and August 3, 2002. The tax exemption is for non-business personal computers and software bundled into one system. The computer is also exempt from gross receipts tax if the purchase price is $1200 or less.

Another approach is the "Income and Bank and Corporation Internet Tax Credit" of California. The legislation encourages the donation of Internet subscriptions and computers by authorizing a tax credit to any taxpayer who is an ISP who provides, without charge, Internet access and personal computers to a low-income household. The amount of the credit for any taxable year would be $200 multiplied by the number of low-income households provided unlimited access to the Internet and computers during the taxable year.

Loan and Grant Initiatives

The enabling legislation to create broadband grant and loans programs vary in budget and target. The majority of grant and loan programs are directed at the public sector. A noted exception is the Arkansas House Bill 2570, which establishes a grant program to eligible telecommunication carriers for extension of facilities to underserved citizens. According to the bill, many areas of Arkansas are still underserved by wireline services. The act establishes an Extension of Facilities Fund to be administered by the Arkansas Public Service Commission (APSC). The bill gives the APSC the right to require each potential customer to be served by the extension of facilities to pay up to $250 for the extension of the service.

As previously stated, the majority of grant and loan programs focus their assistance on local governments. The states of Florida, Colorado, Minnesota, and Nebraska offer grant programs to rural communities for broadband planning and deployment. In Florida, Senate Bill 1460 provides for the establishment of a Digital Divide Trust Fund within the Office of the Governor. The trust is $3 million with the goal of funding local projects that directly alleviate obstacles to information technology access and high-speed infrastructure improvements.

In November 2000, Governor Bill Owens of Colorado provided nearly $500,000 for rural Colorado towns and counties to plan and implement high-speed telecommunications projects. The appropriation came from their "Beanpole" telecommunications grant program that connects rural communities to affordable 21st Century telecommunications systems.

While Florida and Colorado place grant funds in the hands of the governor's office, the state of Minnesota houses their grant-funding program within the Office of Trade and Economic Development. The 2001 Minnesota bill, HF 114, proposes catalyst grants to promote Internet access in rural Minnesota. It allows for the Commissioner of Trade and Economic Development to offer grant funds to local governments and recognized Indian tribal governments to expand broadband Internet access, both fiber-optic and wireless systems. Eligible capital expenditures include equipment and construction costs, but do not include the costs of planning, engineering, or preliminary design. Preference for the grant funding is given to projects that provide access to both business and residential service. The maximum amount for a catalyst grant is $250,000 or 25 percent of the eligible capital expenditures.

The state of Pennsylvania also administers grant funding for technology dispersion through the state economic development office. Introduced in the 2001 session as the "Pennsylvania Community Technology Centers Act," the legislation provides grants and tax credits to establish or expand community technology centers. The technology centers allow disadvantaged residents in urban and rural communities access to information technology. The legislation calls for $5 million appropriation to the Department of Community and Economic Development to fund new or expand existing technology centers. It also provides $2 million for a "New Market Initiative Grant Program" to assist communities in obtaining broadband Internet connection. The bill includes a tax credit for businesses that donate money, equipment, etc., to the centers.

Another approach of note is the funding of rural pilot projects for broadband infrastructure. The state of New Mexico offers comprehensive legislation, entitled the "Telecommunications Infrastructure Fund Act of 2000," that establishes an assistance program to aid local communities in providing telecommunications infrastructure. Under the act, a "Telecommunications Infrastructure Fund Council" reviews proposals submitted by state political subdivisions and community-based organizations to allocate funding from several sources, including severance tax bonds, community development block grants, proceeds from the private use of state and local telecommunications facilities, and other sources to finance needed local telecommunication infrastructure improvements projects. Included in the bill is funding for five demonstration projects in the first year, with each project eligible for up to $1 million in funding. Once the project is completed, ownership of the infrastructure belongs to the local governmental body to be used for "any lawful purpose."

A final grant and loan strategy is the designation of Universal Service Funds for rural broadband access. In Nebraska, Legislative Bill 89 assists rural communities and counties by providing state support for broadband delivery and installation. The bill earmarks one percent of the Nebraska Universal Service Fund to go for grants to small cities and sparsely populated counties to bring in broadband services. It requires a 25 percent match from the local area requesting broadband and ranks applicants based upon high cost factors for providing the service, i.e. "population scarcity and location remoteness."

Broadband Industry Related Legislation

In addition to the laws established at the Federal level with regard to telecommunications and Internet availability, states are also concerned about insuring fair competition among broadband suppliers. It appears from the legislation that a systematic approach exists.

The first is the designation of broadband telecommunications and Internet accessibility as "normal" or "accepted" service. In North Carolina, Senate Bill 641 of 2001 authorizes the North Carolina Utilities Commission to adopt rules expanding the definition of universal service to include statewide Internet access and other technological telecommunications advances. The updated definition of universal service reads as follows:

> In adopting rules to establish an appropriate definition of universal Service, the Commission shall consider evolving trends in telecommunications services and the need for consumers to have access to high speed Telecommunications networks, the Internet, and other services to the extent that those services provide social benefits to the public at a reasonable cost (State of North Carolina 2001).

The second policy approach is the restatement of federal telecommunication legislation at the state level. The wording of the legislation follows and cites federal law and statutes with regard to prohibiting a monopolistic marketplace and ensuring open access for ISPs to consumers. In Hawaii, House Bill 1617 supports the federal position by ensuring competition in the market for broadband Internet access. The legislation notes the importance of a competitive market for offering broadband Internet to the public and prohibits the monopolizing of cable lines. The legislation cites a recent decision of the U.S. Court of Appeals for the Ninth Circuit where they ruled that the transmission of Internet service to subscribers over cable broadband facilities is a telecommunications service under the Telecommunications Act of 1996. Thus, the law permits equal access to broadband Internet and provides for penalties. The act reads:

> Each wireline broadband Internet access transport provider who is, or is an affiliate of, an Internet service provider shall provide any other requesting Internet service provider access to its broadband Internet access transport services, unbundled from the provision of content, on rates, terms, and conditions that are at least as favorable as those on which it provides the access to itself, to its affiliate, or any other person (State of Hawaii 2001).

The state of Hawaii also proposed House Bill 465, which ensures that cable companies provide all ISPs non-discriminatory access to their high-speed cable networks at cost-based rates. The legislation guarantees that Internet access remains affordable and available to all, ensures that cable companies are compensated by ISPs for network usage, and ensures equal access for ISPs.

Similar legislation is offered in Illinois to address high-speed Internet access. House Bill 0138, known as "The Internet Access Enhancement Act," declares that in the best interest of the people of Illinois, no high-speed Internet access transport provider with an affiliated Internet service provider shall have a monopoly. The act guarantees ISPs access to lines held by other ISPs at a reasonable rate. The act notes exceptions to the rule where broadband Internet transport is not offered to the general public. Among the exceptions are federal, state, or local governmental entities, institutions of higher education, business specific fiber, and real estate owners who provide fiber optics to tenants.

In addition to the detailed legislation presented for Hawaii and Illinois, other states passed laws addressing issues of open access and fair pricing with regard to broadband Internet and telecommunications. These states include Washington (House Bill 1077), Rhode Island (House Bill 5235), Oregon (Senate Bill 9), and Virginia (Senate Bill 707/House Bill 1480).

Though private suppliers are the accepted sources of broadband accessibility, states offer broadband opportunities for alternative providers to enter the market because of the failure of private enterprises to properly service rural communities. Several pieces of legislation exist in states with large rural populations. In California, Assembly Bill 954 allows irrigation districts to purchase or lease telecommunication services. In recognition of the importance of broadband telecommunications for rural economic development and the lack of availability of broadband in certain rural areas, irrigation districts with existing broadband infrastructure can lease excess capacity to businesses and residents within the boundaries of their region. It also allows irrigation districts to purchase or lease telecommunications equipment from any agency or entity, public or private, to distribute telecommunications services to the proposed areas.

In 2001, a similar piece of legislation directed toward rural communities was offered in Tennessee. Senate Bill 190/Senate Bill 977 authorizes municipal electric suppliers to provide cable, Internet access, or related services outside its service area, but within the county boundaries where the system is principally located. The expansion of service authority applies only to systems with less than 4,000 residential accounts and in counties with a population of less than 40,000.

Though various versions of cooperatives are accepted among private companies as providers of broadband, serious objections have arisen with states empowering government entities to own fiber. In Nebraska, Legislative Bill 827 introduced by Senator Bromm on January 17, 2001, allows government entities to own, sell, or lease "dark fiber." "Dark fiber" refers to unused fiber optic cable where no light is transmitted or is without a signal. Yet, the bill does not allow governmental agencies to provide telecommunications services. If a governmental agency sells or leases their dark fiber, profits earned from the transaction are remitted to a newly created "Nebraska Internet Enhancement Fund." The legislation ignited objections from the state telecommunications industry. The Lincoln Electric System (LES), a city-owned utility, sought to run 70 miles of dark fiber around the city to offer business customers broadband data services. According to a report from the Nebraska State Paper.com (2000), Jack Schultz, attorney for the Nebraska Telecommunications Association, argued that a city-owned public power utility should not be authorized to run dark fiber because "it (LES) wants to pick and choose whom it serves." In response, Mark Ayotte noted that Lincoln is one of only two home rule charger cities, which gives it broad authority to do most anything it wants unless it runs afoul of constitutional matters. Thus, Ayotte presented the LES example as a unique situation, not something of statewide commercial importance.

In order to prevent conflict among private providers and government entities, the state of Virginia, passed legislation in 1999 prohibiting localities from offering telecommunications services to consumers. The legislation, Chapter 916, states:

> No locality shall establish any department, office, board, commission, agency, or other governmental division or entity which has authority to offer telecom-

> munications equipment, infrastructure, other than pole attachments or tower attachments including antennas or conduit occupancy, or services, other than intragovernmental radio dispatch or paging systems shared by adjoining localities (State of Virginia 1999).

The legislation provides for a locality, electric commission, or industrial/economic development authority to lease "dark fiber." Under the provision, no profit can be made by the local agency leasing the fiber; only the cost of installing and maintaining the fiber can be recouped through the lease agreement.

In Idaho, lawmakers went a step further with House Bill 228 in 2005. HB 228 establishes a minimum for telecommunication service quality. The minimum standards were needed as the state transitioned from a regulated market to open competition to ensure that citizens continued to receive high quality telecommunications.

Economic Development Initiatives

Though many policy initiatives emerge at the state level, the economic development initiatives center upon broadband telecommunications for rural communities are quite varied. These legislative proposals range from strategic planning to outreach programs. This section of the chapter suggests particularly great potential for the development of policies that assist rural communities in acquiring and utilizing broadband telecommunications for job creation.

In order to understand the full depth of the situation, Colorado examines broadband technology from a geographical and economic development perspective. House Joint Resolution 01-1053 calls for the House Information Technology Committee and the Joint Legislative Computer Management Committee to study, receive testimony, and make recommendations on communications and information technology issues affecting state government and to propose legislation to address issues identified by the committee. Study topics include: a statewide strategic plan that identifies the state's information technology infrastructure needs, economic development-related technology needs, and statewide broadband capabilities.

In order to plan for broadband-based economic development in rural Idaho, the state created the "Governors Task Force on Rural Development" to inform the executive branch of economic development issues important to rural communities and citizens. One of the first recommendations identified the need for broadband telecommunications infrastructure throughout the state of Idaho. In a report from the Governors Task Force on Rural Development

(2000), the two major recommendations for telecommunications infrastructure included the following:

1) The state should maximize the use of the Idaho Internet infrastructure as a means to leverage the development of affordable broadband telecommunications service to rural areas, private businesses, and public entities.
2) The state of Idaho should immediately explore options and implement actions to promote broadband usage in rural areas.

A similar legislative proposal from North Carolina, Senate Bill 1343 of 1999, created the "North Carolina Rural Internet Access Authority" to study and report on the current condition of information technology infrastructure and identify the IT needs in rural regions of the state. The authority consists of 21 members, six appointed by the Governor, 12 appointed by the General Assembly (six in Senate, six in House), and three ex-officio, voting members: the State's Chief Information Officer, the Chair of the North Carolina Rural Economic Development Center, and the Secretary of Commerce.

In Georgia, Senate Bill 465 establishes the Georgia Technology Authority (GTA), the successor to the GeorgiaNet Authority. The purpose of GTA is to provide for the procurement of technology resources, technology enterprise management, and technology portfolio management for the state of Georgia. Among the many duties of GTA, it is to establish the architecture for state technology infrastructure, promote efficient use of resources, and utilize the system for economic development. On January 12, 2001, Governor Roy Barnes announced that the GTA would launch an initiative to build a state of the art communications network connecting the entire state of Georgia. The new network improves constituent access to government and promotes economic development by providing a world-class infrastructure to every community in Georgia (Office of the Governor of Georgia 2002).

In order to promote rural economic development in Hawaii, the legislature authored a pair of bills offering a traditional and a non-traditional approach to rural broadband access and usage. A more traditional approach, Senate Bill 542, makes blank appropriation for government obligation bonds to extend the community-based economic development loan and grant programs for IT initiatives in distressed rural areas. The legislation noted the success of the community-based economic development loan and grant program and expanded the scope of the plan to reduce the effects of the "digital divide." In contrast to traditional economic development tools, Senate Bill 678 proposes an innovative IT-based program. The bill creates the New Economy Transition (NET) Program for Hawaii. This legislature notes the importance of high-speed Internet in business activities, from marketing to e-commerce, as critical for sustainability for the new information economy. The NET program is developed in the mold of the agricultural extension model. The NET program reaches out to businesses in Hawaii, fostering partnerships and assisting businesses transitioning into the new economy. The legislation notes that businesses will benefit from the NET program because of greater access to technology programs and sources of funding, increased productivity and efficiency, and the ability to compete locally and globally. The government benefits from the NET program through increased participation in the new economy, greater employment opportunities, higher tax revenue, increased global competitiveness, and a more highly skilled workforce.

One of the most unique programs to assist rural economic development is from Iowa. House File 263 provides for the establishment of a rural community Internet access program for businesses in communities lacking affordable, high-speed broadband Internet access. In order to participate in the program, a community, defined as a city, unincorporated area, or township, shall have a population of 20,000 or fewer individuals, possess less than T2 capability, and agree to utilize the Iowa communications network for Internet access strictly for economic development activity by community businesses. The bill also requires that an application must be filed with the Iowa Economic Development Board to participate in the program and includes an assessment of the length of time anticipated before the community has high-speed access. The access is not subsidized by the state of Iowa.

Connecticut takes this one step further with Bill 6706 that prohibits a fee for terminating telecommunication or broadband services to small businesses. The bill provides that any small business (less than fifty full-time employees) cannot be charged a termination fee for terminating any broadband or telecommunications services.

The final pieces of rural economic development legislation are adaptations of existing economic development techniques to create IT-related jobs. Similar to "enterprise zones," the states of Oregon and West Virginia propose "zones" where Internet/technology intense businesses can locate and avoid certain taxes. In Oregon, Senate Bill 229 authorizes the Director of the Economic and Community Development Department to designate three areas as "electronic commerce zones." Businesses locating in the zone would be exempt from certain property taxes and allows additional credit against income or corporate excise tax. While in West Virginia, in a report submitted to Governor Bob Wise, the Technology in State Government Taskforce suggested the creation of an "Internet Free Trade Zone." By creating a zone with high tax relief, they feel there may be a possibility that Internet based companies may consider relocating to West Virginia (O'Hanlon and Comstock 2001).

CONCLUSION

This review of federal and state legislation provides a structural basis for the development of a model for implementing broadband telecommunications for rural economic development. From the analysis of legislative and executive proposals, certain patterns appear in the development of state and federal policy. From the federal legislative perspective, policies are directed toward private firms supplying the broadband telecommunications infrastructure to rural Americans. The proactive federal policies center upon tax cuts for telecommunications investment and low interest loan programs. The most apparent function of federal policy is assuring that competition remains fair among telecommunications companies and cable television providers. Yet the vital ingredient of federal broadband policy is the idea of timeliness of broadband delivery to rural America, now a function of the Federal Communications Commission.

The state legislative and executive policies provide numerous approaches to supplying rural communities with broadband telecommunications. From the analysis of state policy, three trends appear. First, it appears that states that have presented numerous pieces of rural broadband legislation, i.e., Virginia, Washington, and North Carolina, first commissioned studies on the standing of telecommunications infrastructure within each respective state. Secondly, states that have developed comprehensive rural broadband strategies tend to have a large metropolitan area within the state, i.e., Washington, Georgia, and Virginia. The presence of a large metropolitan area market could serve as a way to leverage private telecommunications companies to provide broadband rural service. Third, it is evident that a "policy divide" exists between states that have built upon the opportunities of the digital economy and those who have not. For instance, states such as Virginia and North Carolina have aggressively passed legislation and initiated programs to further expand digital communications technologies for rural citizens, where extremely rural states, such as North Dakota and Mississippi, have failed to do so.

In conclusion, based upon the review of legislation, predominantly rural states are not only falling behind in broadband telecommunications service but also in the development of telecommunications related policy. It appears that the fastest way for states seeking to address broadband telecommunication policy deficiency is to construct a comprehensive rural broadband policy package that draws from successful examples from more progressive states. Based on the policy initiatives of states examined in this chapter, the policy package should begin with an inventory of broadband telecommunication infrastructure within the state and followed by a package of innovative strategies, i.e., broadband tax credits, grant and loan programs, to encourage broadband deployment and consumer utilization.

Chapter Four

Beyond Access: A Cluster Approach to Broadband Telecommunications-based Rural Economic Development Planning

In recent years, issues such as the "digital divide" and "broadband telecommunications access" have been of particular importance to both economic development professionals and academicians. The increasing competitiveness of the global economy requires economic development professionals to understand the condition of technological infrastructure, particularly telecommunications, within their respective communities (Kolzow and Pinero 2001; Wohlbruck and Levy 2001).

The attention given to technology-based development strategies by economic development professionals illustrates the significance of the issue. The importance of technology-based economic development is demonstrated by national, regional, and state professional economic development member organizations, such as the International Economic Development Council (IEDC), the Southern Economic Development Council (SEDC), and the Mississippi Economic Development Council (MEDC), respectively. Each of these organizations addresses the value of telecommunications-based economic development strategies through continuing education classes, conference sessions and roundtable discussions, and publications, such as the *Economic Development Review* and *SEDC News*. Beyond the recognition by professional economic development organizations, programs incorporating telecommunications capabilities with local economic development practice exist. These economic development strategic plans center on the application of information technologies, including broadband telecommunications, for local economic development. This type of technology focused economic development strategic planning is utilized in numerous states, such as Arizona and Georgia (Arizona Department of Commerce 2002; Georgia Department of Industry Trade and Tourism 2002).

Telecommunications-based economic development is significant to researchers because of its effect on business structure and organization. In addition, researchers are interested in the potential of telecommunications to alter the importance of location in economic development activity. Much of the academic literature dealing with local and regional economic development in the United States is grounded in the Weberian theory of industrial location (Dicken and Lloyd 1990). This theory based on the work of Alfred Weber (1909), uses transportation costs as the key factor in industrial location decisions, resulting in agglomeration economies. But as the industrial base of the United States shifted from manufacturing to services, becoming less dependent upon the movement of physical goods, new challenges emerged for academicians to develop applicable theories that explain locational significance with regard to the movement of information. According to theorist Manuel Castells, emerging technologies, such as telecommunications, have changed the applicability of Weberian-based economic development strategies:

> The most direct impact of high-technology on the spatial structure concerns the emergence of a new space of production as a result of two fundamental processes: on one hand, high technology activities become the engine of economic growth and play a major role in the rise and decline of regions and metropolitan areas, according to their suitability to the requirements of high-tech production; on the other hand, the introduction of new technologies in all kinds of economic activities allows the transformation of their locational behavior, overcoming the need for spatial contiguity (Castells 1985, 12).

As a result of the decline of manufacturing as the leading industrial sector within the United States, practitioners and academics adopted new strategies and theories to address economic development within the context of the emerging global, information economy. One such theory receiving significant attention is the "Theory of Competitive Advantage," within Porter's *The Competitive Advantage of Nations* (1990), commonly referred to in practice as "clusters."

According to Burress (2001, 1), clusters are a "hot topic" in the world of economic development, with many states and regions using it as the basis for economic development strategic planning. Though without a standardized definition, clusters generally are defined as "businesses in related industries that are linked in some way and that tend to co-locate in geographic space" (Malizia and Feser 1999, 226). Defined in this manner, clusters suggest a group of industries joined together through backward or forward linkages and located within a geographic space, allowing for frequent interaction. Thus, based on the definition, cluster-based economic development planning centers on locations that possess existing strengths within certain industrial sectors to further promote its competitive advantage. Therefore, a cluster strategy of "building upon linkages within a geographic area," as suggested by Malizia and Feser (1999, 227), would generally focus upon the urban areas within states and regions.

Despite the potential for urban bias within cluster strategies for economic development, these approaches are readily adopted by states with high rural populations. For example, South Carolina's economic development plan is a cluster strategy intended to jumpstart five or six key industries in which the state already has a head start (Clemson University Community and Economic Development Program 1996). Thus, based upon its inherent urban-biased structure, how does cluster-based planning address the economic development concerns of rural areas?

Limited research has been conducted to identify rural areas that have the potential of using clusters for economic development. Though without the intricate, and sometimes complex, linkages associated with industrial clusters, some rural communities are building on the "clusters of knowledge" model for economic development. The idea of "rural knowledge clusters" is a research initiative of the Hubert H. Humphrey Institute of Public Affairs at The University of Minnesota. According to this research, rural knowledge clusters are "innovative, interrelated groups of firms outside metropolitan regions, deriving competitive advantage through accumulated, embedded, and imported knowledge among local actors and institutions; these can be explained in three factors: competitive advantage, history, and institutions" (Munnich, et al. 2002, 7). Within this framework, "rural knowledge clusters" are developed as a model for encouraging innovation in rural areas, depending heavily upon information technologies, in particular broadband telecommunications.

Although "rural knowledge clusters" represent a new approach to rural economic development planning based upon information technologies, there are numerous examples of programs and policies that examine the potential role of broadband telecommunications for economic development. These "experimental" programs include such initiatives as "Intelligent Communities" and "Information Age towns" (World Teleport Association 2002; eircom 1996). These noted programs, in the communities of LaGrange, Georgia and Ennis, Ireland, respectively, validate the idea that jobs can be created based upon the implementation of information technologies.

Another innovative program that follows in the same form is the Centers for Excellence in Rural America (CERA) Program of the Western Governors' Association. The purpose of the CERA program is to test the hypothesis that creating a network of small rural towns deploying affordable, high-speed telecommunications services will result in increased job creation; with the significant assumption being that these technologies will alleviate the distance barrier experienced by their physical location (Western Governors' Association 1999). The CERA project is driven by the question: How can rural communities utilize broadband telecommunications for job creation?

PURPOSE OF THE CHAPTER

The central question of this chapter is the potential for rural economic development based upon broadband telecommunications. This question is addressed through the theoretical framework of competitive advantage and clusters, using the communities of the Centers for Excellence in Rural America (CERA) program as the focus areas of study.

It is widely accepted that access to broadband telecommunications alone is not sufficient for successful rural economic development (Korsching, et al. 2000). Thus, the cluster theory framework suggests the following approach: "Beyond access, what factors are important for rural economic development based upon information technologies and broadband telecommunications?"

This research is conducted under the assumption that "sufficient" broadband telecommunications are available in every county of the study area. "Sufficient" broadband telecommunications are defined as DSL capabilities or beyond, more specifically transmission speeds of 1.5 million bps or greater, in both directions.

The study areas are the rural counties of North Dakota and Wyoming, including selected border counties in South Dakota, Minnesota, Montana, and Nebraska; 96 total counties (Figure 4-1). "Rural" counties for the research were determined by using the standard definition of the U.S. Bureau of the Census, in particular, counties with a population per square mile of 15 persons or less. The rural counties in the study were selected because: 1) the upper Midwest is the most rural region of the U.S.; and 2) they contain all of the communities involved in the CERA program. Five counties contain member communities of the CERA project. Known as CERAtowns, they are located in: Traill County, North Dakota; McKenzie County, North Dakota; Niobrara County, Wyoming; Park County, Wyoming; and Converse County, Wyoming.

RESEARCH METHODOLOGY

Factor analysis is the statistical technique used to determine which variables are the most significant in the use of broadband telecommunications for rural economic development within the study area. The data sources for the statistical analysis are from the Bureau of the Census, Bureau of Labor Statistics,

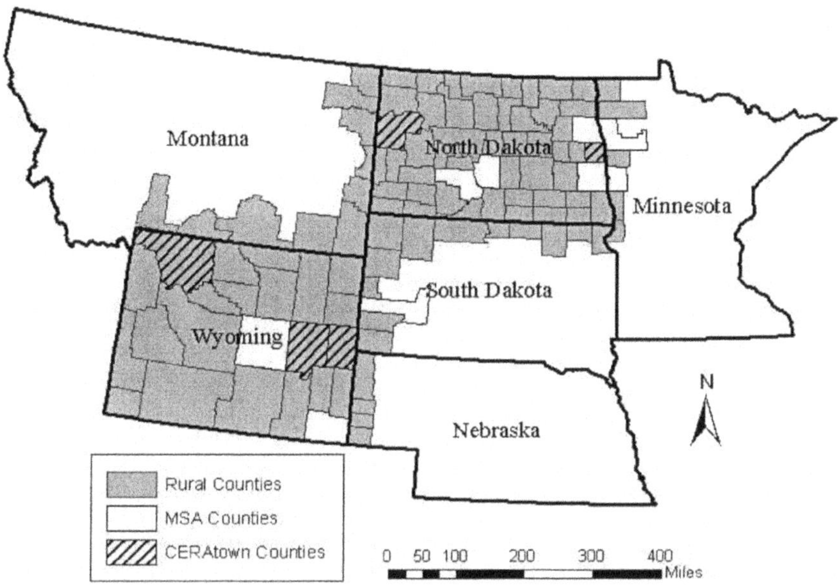

Figure 4-1. Areas of Study.

the State of North Dakota, the State of Wyoming, the State of Montana, the State of Minnesota, the State of South Dakota, and the State of Nebraska for the years 1990–2000. The results of the factor analysis are used to determine which of the CERAtown communities are most likely to be "successful" in implementing high-speed telecommunications technology for economic development by producing a "broadband development index" based upon the factors emerging from the analysis.

The research is presented in the following sections: 1) A Theoretical Summary of Competitive Advantage and Cluster Analysis and Its Application for Rural Economic Development; 2) The Centers of Excellence in Rural America (CERA) Program and Member Communities; 3) Research Methodology and Data; 4) Results from the Factor Analysis; 5) The "Broadband Development Index," and 6) Conclusion.

A THEORETICAL SUMMARY OF COMPETITIVE ADVANTAGE AND CLUSTER ANALYSIS AND ITS APPLICATION FOR RURAL ECONOMIC DEVELOPMENT

Many of the current trends in the area of economic development policy are strategies based upon the theory of competitive advantage or clusters (Rosenfeld

2001). The theory of competitive advantage, or cluster theory, is commonly associated with the research work of Michael Porter and his book, *The Competitive Advantage of Nations* (1990). Despite the great attention clusters have received over the last decade, it is not a new phenomenon. According to Malecki (1997a, 9), much of the work regarding clusters and economic competitiveness developed by Porter is grounded in the "concerns of neoclassical analysis and Schumpeterian variants theories" and the work of theoretical economic geographers. According to Rosenfeld, a cluster is:

> A geographically bounded concentration of similar, related, or complementary businesses with active channels for business transactions, communications and dialogue, that share specialized infrastructure, labor markets, and services, and faced with common opportunities or threats (Rosenfeld 1995, 23).

The design of cluster development is embodied within the theory of competitive advantage. The foundations of Porter's theory are the determinants of competitive advantage, presented in the "diamond" model of competitiveness (Figure 4-2). The "diamond" consists of four key determinants: Factor Con-

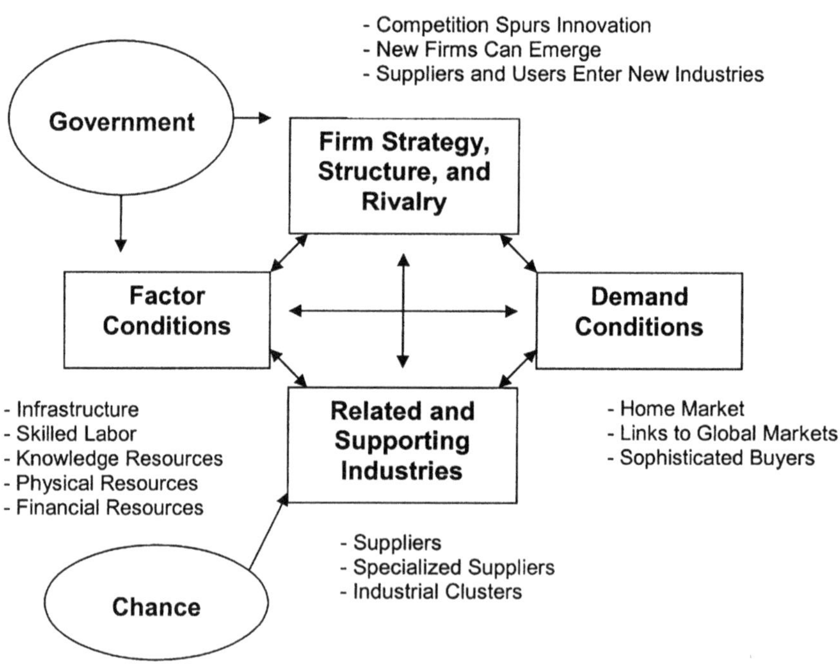

Figure 4-2. The Complete "Diamond" of Competitive Advantage Structure. *Sources:* (Porter 1990, 72; Malecki 1997, 8).

ditions, Related and Supporting Industries, Demand Conditions, and Firm Strategy, Structure, and Rivalry. In addition to the four key determinants, the variables of chance and government are included "to complete the theory" (Porter 1990, 73). Factor conditions include the traditional inputs of industrial production, such as land, labor, capital, and knowledge/entrepreneurship, commonly termed the factors of production. Also included within the category of factor conditions are two additional factors, sophisticated knowledge institutions and advanced infrastructure.

Demand conditions refer to the demand for the products within the area of the home market. As noted by Porter (2000), areas with sophisticated buyers tend to reinforce the magnitude of the clustering capabilities of a location. Related and supporting industries are suppliers who share information, skilled labor, and other services to steadily build the level of excellence in the products or services created. Firm strategy, structure, and rivalry are the "legal, cultural, and institutional framework" that determines the competitive position of firms within a country, region, or place (Malecki 1997a, 9).

The final pieces, government and chance, represent external forces potentially influencing any of the four major determinants of competitive advantage, making a cluster greater in strength or weakening its position.

The framework of the "diamond" model allows for numerous applications within the field of economic development. The components of the model apply to different scales of economic structure, in particular: the micro level (firms), meso level (locality or region), and macro level (national) (Willoughby 2001). The micro level studies focus on the competitiveness of individual firms, analyzing how a firm is positioned within a particular industry (Gibbs and Bernat 1997). The meso level applications of cluster analysis examine the competitiveness of regions within various industrial clusters. An example of meso level inquiry is the potential role of university research parks in the development of regional clusters, such as the Ohio Aerospace Institute (U.S. Economic Development Administration 1997). As with Porter's use of international case studies, macro level inquiry addresses the competitive stance of nation versus nation.

Despite the numerous levels of analysis, rural communities receive little attention, both in terms of potential cluster development ability and the effect of clusters on their development potential. According to Burress (2001), the use of clusters for economic development planning clearly is biased toward urban areas. With many levels of government concerned about funding for economic development activity, cluster strategies offer "the most bang for the buck" (Burress 2001, 2). But this approach could be potentially dangerous for rural communities that essentially become insignificant in state-level development plans.

Although the majority of cluster analysis and competitive advantage research examines nation, state, and urban levels, research applications using competitive advantage in rural communities as a framework of analysis do appear (Isserman 2001; Drabenstott 2001). Based on the literature, rural cluster research may be categorized into two distinct themes: industrial clusters and innovation clusters. Typical studies addressing cluster development in rural areas examine industry-specific examples of rural competitive advantage, such as the carpet industry in Dalton, Georgia; auto supply chains in Tennessee; houseboat manufacturing in Kentucky; and furniture manufacturing in Tupelo, Mississippi (Rosenfeld, et al. 2000). While in contrast to industry specific clusters, innovation studies address the potential of rural regions to build clusters based on components of knowledge and institutions (Isserman 2001; Jarboe 2001). Within the limited literature on the topic, the most significant work in the field of rural innovation, referred to as "rural knowledge clusters," is currently conducted at the Hubert H. Humphrey Institute of Public Affairs at The University of Minnesota.

The Humphrey Institute research is grounded in the theoretical framework of "rural knowledge clusters," developed from Porter (1990). The reason for using "knowledge" as the foundational point of this research is because competitive advantage in the global information economy is dependent upon it (Porter 1998). According to The Humphrey Institute, the model of "rural knowledge clusters" is based on three "elements of dynamism" for a given rural location: competitive advantage, historical development, and institutions (Munnich, et al. 2002). Competitive advantage, in this model, refers to beneficial economic development factors, such as a rich base of skilled workers, access to proximate market opportunities, or local entrepreneurial culture. Historical development is the unique business or educational structure in a community that creates knowledge, noting that knowledge does not appear out of thin air. Institutions, both formal and informal, are instruments of education that foster the creation, diffusion, and renewal of the local knowledge base. According to Munnich, et al. (2002), these "institutions" are usually formal in nature, such as community colleges and universities. The theoretical framework is presented in Figure 4-3.

For this research, the theory of "rural knowledge clusters" is used as a framework to develop an applied model for rural economic development planning. As noted earlier in the manuscript, the driving forces behind the "new" economy are knowledge and technology. Thus, competitive advantage in the information economy occurs in locations that possess skilled labor and advanced technological infrastructure, which includes broadband telecommunications. Figure 4-4 is a modified version of the "rural knowledge clusters" theoretical framework that represents how this study uses its structure to develop

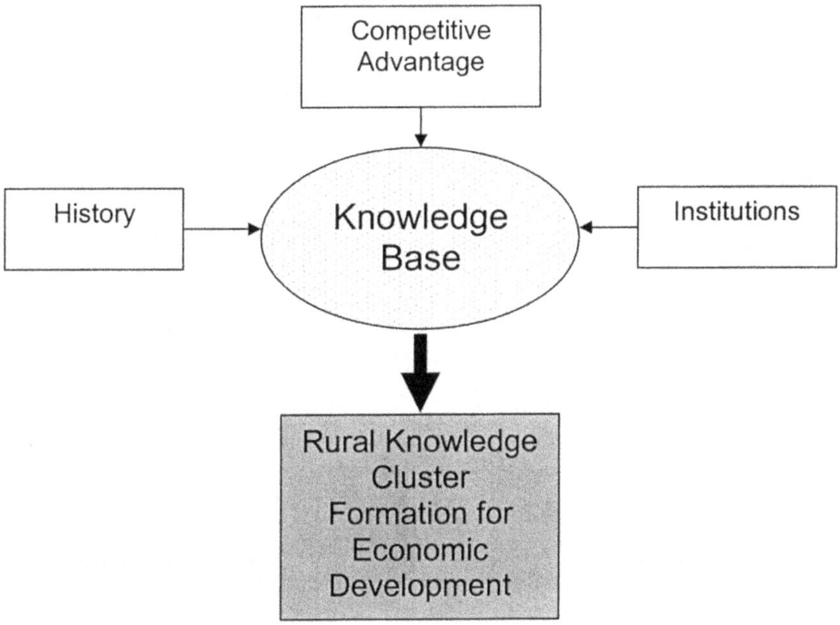

Figure 4-3. Rural Knowledge Cluster Framework *Source:* Munnich, et al. 2002.

an applied model for rural economic development planning. This research focuses on the following dynamics important to technology-based economic development: leadership, location, labor, economic and social characteristics, higher education and research institutions, and technological capacity.

In order to develop the model, it is necessary to examine these dynamics in more detail. Therefore, using the CERA program as a basis of study enables this research to determine the significance of these dynamics.

THE CENTERS OF EXCELLENCE IN RURAL AMERICA (CERA) PROGRAM AND MEMBER COMMUNITIES

The Centers for Excellence in Rural America (CERA) Program, an initiative of the Western Governors' Association with leadership from the Governors of North Dakota and Wyoming, was created to enhance the potential economic development opportunities of rural communities through advances in telecommunications and technology. Formed in 1997, CERA is a five-year program to test the hypothesis that creating a network of small rural towns deploying affordable, high-speed telecommunications services will result in increased job creation—with the significant assumption being that these

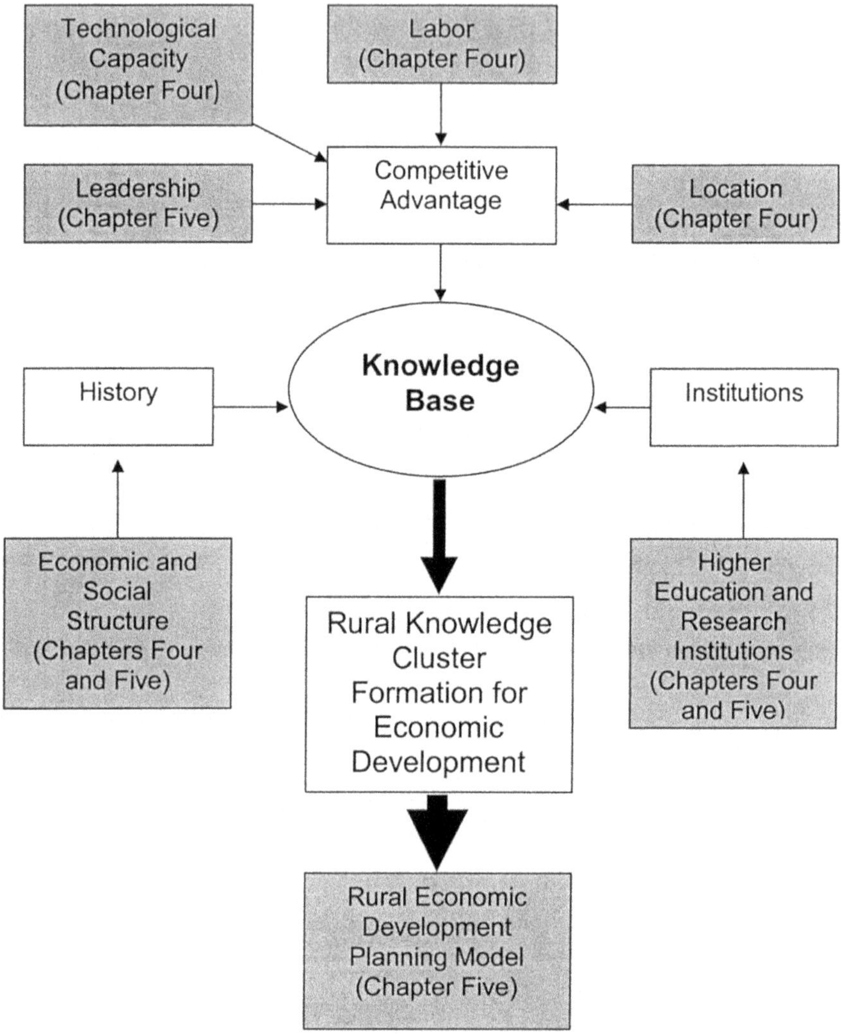

Figure 4-4. Author's Adaptation of Rural Knowledge Cluster Framework. *Source:* Munnich, et al. 2002.

technologies will alleviate the distance barrier experienced by their physical location (Western Governors' Association 1999). From the outset, three primary goals existed for the CERA program:

1) To demonstrate that broadband telecommunications can enhance rural communities by providing better jobs, enhanced health-care access, education, and government services

2) To demonstrate that rural communities can maintain their heritage and quality of life, yet be viable competitors in the international marketplace
3) To demonstrate a virtual community model that may be replicated in other communities in rural America

The goals of the program are achieved through the development and testing of "pilot" communities, known as "CERAtowns." The CERAtown communities would "research, identify, and implement information age technologies in six application areas: Earning a Living, E-Commerce, Wellness, Education, Citizen Participation, and Leisure" (Western Governors' Association 1999, 6).

The CERA program operates at three levels: CERA USA, CERA states, and CERAtowns (Figure 4-5). CERA USA serves as the governing body of the program through representation from each of the CERA states and CERAtowns. Because of the membership level of the Western Governors' Association, the CERA program has a potential membership capacity of 17 states. According to the operational plan of the CERA project, CERA USA performs the following functions under the auspices of the WGA: overall project management and coordination, financial management and reporting, corporate partner relations, overall information exchange and coordination, funding strategy for the CERA program, development of guidelines for expansion, and development of criteria to be met by CERAtowns. CERA states are represented by a State Coordinator appointed by the Governor of that state who serves as an intermediary between CERA USA and the Governor. CERAtowns represent the "test sites," where the groundwork is laid for the model development of the overall CERA program. The duties of the CERA-town include: local leadership and coordination, creation of an information

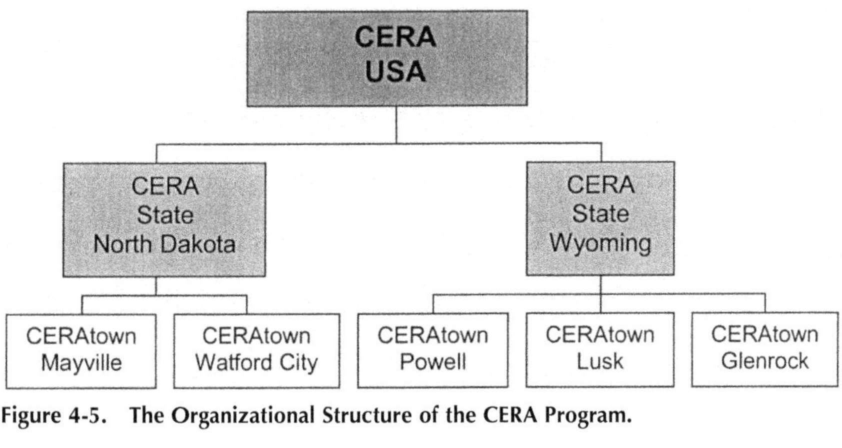

Figure 4-5. The Organizational Structure of the CERA Program.

technology strategic plan, provide facilities for the technology network center, provide necessary community staffing support, be actively involved in the CERA project and share results, mentor other towns, and monitor performance metrics for CERA.

The original funding proposal for the five-year project was nearly $16 million, with the majority of the financial support from federal sources ($11.5 million). In the first year, project work centered upon initial program set-up duties, facilities and utilities, and community infrastructure assessments projects within each of the CERAtowns. The second year included the addition of sales and marketing personnel to maintain the offices, as well as actively recruiting business for location with the CERAtowns. The third through fifth years meant an increasing dependence upon local resources and the carrying out of evaluation studies to identify CERAtown successes and failures for the development of the comprehensive model.

The original communities selected during year one of the project included two from North Dakota and two from Wyoming. According to the CERA plan, the primary site selection criteria for the pilot sites were a population of at least 1,000, but less than 10,000, and local community leadership needed to demonstrate an understanding and desire to participate in the CERA program. Based on these standards, the following communities were selected: Watford City, North Dakota (McKenzie County), Mayville, North Dakota (Traill County), Powell, Wyoming (Park County), and Lusk, Wyoming (Niobrara County). In 1999, CERA added the community of Glenrock, Wyoming (Converse County) during the third year of the project. Figure 4-6 presents the geographic locations of the participating CERA communities.

Watford City, North Dakota is a ranching and oil community located in McKenzie County, just north of Theodore Roosevelt National Park in western North Dakota. The population of Watford City is 1,435 (U.S. Census 2000). Before the CERAtown project, Watford City was one of the first communities in North Dakota to develop a "Community Technology Plan" that acknowledged the difficulties of attracting high-tech industries with inadequate telecommunications infrastructure (Western Governors' Association 1999). The CERA trial project for the community is the construction of a community-based technology and education/training center, utilizing the high-speed telecommunications capabilities of the local high school.

Mayville, North Dakota, in Traill County, is a farming community and home to Mayville State University located in the Red River Valley of eastern North Dakota. The population of Mayville is 1,953 (U.S. Census 2000). The CERA trial project for the community is the establishment of a business incubator, housed at Mayville State University, to support the creation of information technology dependent businesses. In 1998, Mayville State received

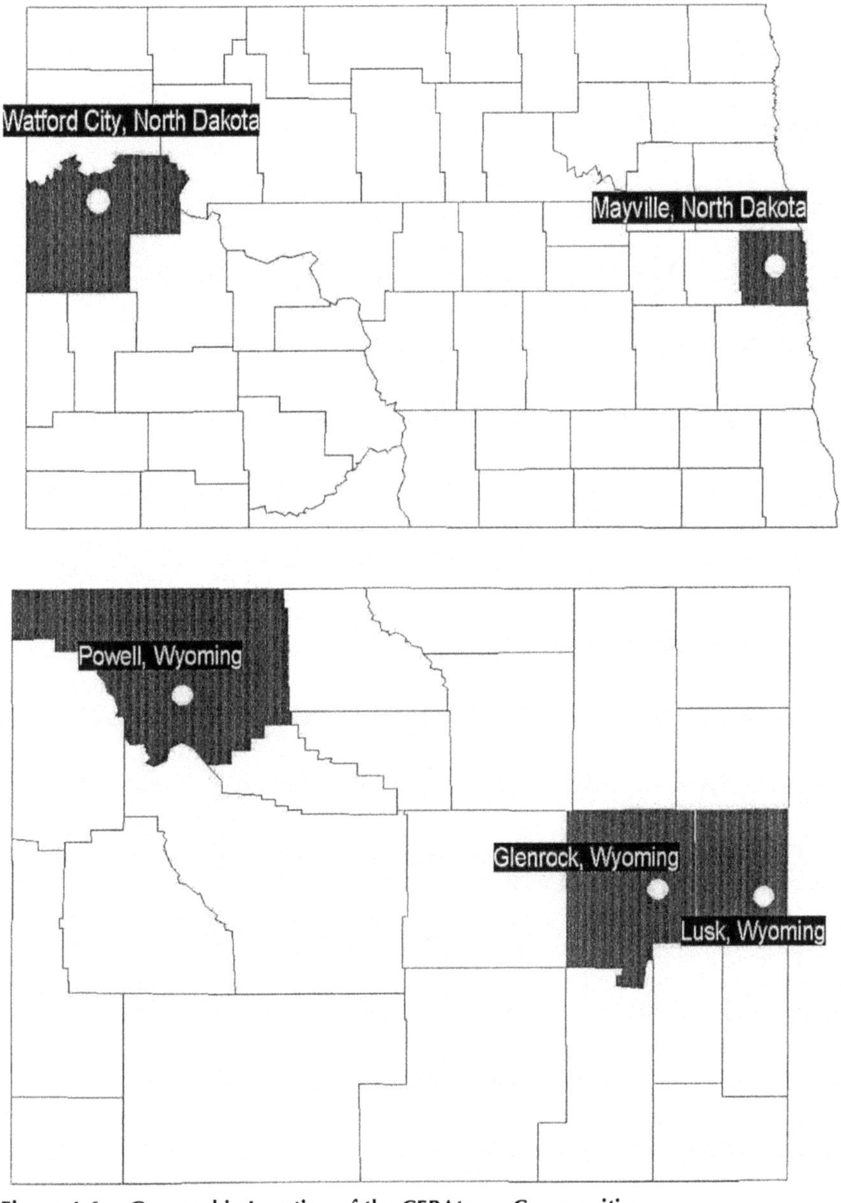

Figure 4-6. Geographic Location of the CERAtown Communities.

a Department of Commerce Technology Opportunities Program (TOP) grant to formally establish the incubator and support its daily activities, including staff and facilities management (Traill County Technology Center 2002).

Lusk, Wyoming is a ranching community located in east central portion of the state. Lusk has a population of 1,649 (U.S. Census 2000). The CERA initiative of the community focuses upon the creation of home-based businesses. The City of Lusk provides broadband access to every home within its limits and the local CERA representatives offer assistance and training to individuals wishing to utilize their connectivity for the development of Internet-based businesses.

Powell, Wyoming is a farming and college community east of Yellowstone National Park. Powell is the largest of the CERAtown communities with a population of 5,687 (U.S. Census 2000). The Powell project in the CERA program is the creation of an "Innovation Network Center" located at Northwest College. The Center provides a combination of technology-based services including a business technology incubator, a telecommuter workstation center, technology-related training, and e-commerce development service for local businesses. Besides the CERA project, the same facility at Northwest College is home to both the field office of the Wyoming Business Council and the Northwest College Workforce Training and Development Program.

Glenrock, Wyoming is located in central Wyoming at the foot of the Rocky Mountains, where Deer Creek converges with the Platte River. Glenrock has a population of 2,307 (U.S. Census 2000). As a late entry in the CERA program, Glenrock failed to establish any substantial initiatives before the end of the program in 2001.

Despite the potential promise of the CERA program, the comprehensive funding from the federal government did not take place. Though without significant financial support, the member CERAtowns followed the structure of the program with resources from the Department of Commerce, the Western Governors' Association, the State of North Dakota, the State of Wyoming, and local funds from city and county government. Though they could not fund the entire project, the participating communities continued to meet and share information about economic development opportunities and strategies.

Using CERAtowns as a basis of study is important to the research because the test communities provide the opportunity to conduct follow-up interviews with regard to broadband effectiveness for economic development. Because of their involvement in the project, each CERAtown possesses an increased awareness with regard to the application of broadband telecommunications for economic development. The following traits are important to the research and exist in the CERAtown communities: they have broadband capabilities, they have developed tangible projects based on advanced telecommunications, and they use information technology as a component of their local economic development strategy.

ANALYTICAL APPROACH

As established in Chapter Two, broadband telecommunications is a critical element for remaining competitive within the global, information economy. It is also recognized that broadband telecommunications policies, both at the federal and state levels, focus on the achievement of access for rural areas. Yet, these policies do not effectively address the most critical issue of how to use broadband telecommunications for economic development. According to Korsching, broadband is not the only provision for successful rural economic development.

> A growing body of evidence indicates a significant positive relationship between investments in telecommunications and economic growth . . . however, our research findings and others demonstrate that telecommunications technologies alone are not sufficient for rural economic development, and blind pursuit of their benefits could result in significant unintended or unwanted results (Korsching, et al. 2000, 280).

Thus, telecommunications-based development, just as with any other traditional economic development strategy, still operates in an environment where not all locations are created equal. Endogenous traits, such as interstate access, universities, access to rivers, skilled labor, etc., make the economic development situation for some communities more advantageous than others. Although broadband telecommunications are a crucial component for the efficient movement of information in the global economy, not every location is equal in its ability to use it. So what happens when every rural location in the United States has access to broadband telecommunications? Who will benefit economically?

The purpose of this chapter is to explore the potential of rural communities to utilize the broadband telecommunications for economic development, moving beyond access to implementation. The research is carried out through the following steps:

1) Identify the variables most important to the use of broadband telecommunications through the work of previous research studies
2) Use factor analysis to identify the statistically relevant variables and reduce the identified variables to manageable set of factors for the study area

3) Produce a "broadband development index" that ranks the study area counties based on the emerging factors, revealing the potential of counties to utilize broadband telecommunications for economic development
4) Use Geographic Information Systems (GIS) to identify spatial patterns for the factor scores and broadband development index
5) Identify the CERA counties within the index and use their rankings to determine which locations are used for the follow-up qualitative research in Chapter Five

Once again, this research is conducted under the assumption that all of the counties within the study area have "sufficient" broadband telecommunications. Through this assumption, the potential for competitive advantage gained through access versus non-access is eliminated—thus, allowing the analysis to focus upon variables that influence the use of information technologies and broadband telecommunications.

As established earlier in the chapter, the theoretical framework of "rural knowledge clusters" focuses on three core components: competitive advantage, history, and institutions. The University of Minnesota uses case-study techniques to identify variables important to the establishment of "rural knowledge clusters." Some factors identified by Munnich, et al. (2002) as important for the development of "rural knowledge clusters" are: colleges and universities, leadership, local economic structure, and entrepreneurship. While this chapter uses quantitative methods, non-quantifiable factors, such as local leadership and entrepreneurship, are addressed in Chapter Five.

The analytical tool used for this research is factor analysis. It is a quantitative method used to analyze the structure of interrelationships (correlations) among a large number of variables by defining a set of common underlying dimensions known as factors (Hair, et al. 1998). This method of analysis is a common quantitative method in numerous fields of academic research, including economic geography and sociology (Coulson and Darling 1996). Specifically, factor analysis is a commonly applied tool for cluster research (Sung and Gibson 1998; Young, et al. 1993; Neiman and Fernandez 1999). Due to the extensive use and acceptance of this statistical technique, a detailed explanation is not presented, but the interested reader may refer to the book *Multivariate Data Analysis* (Hair, et al. 1998) for basic methodological information.

Existing data sources are used to conduct the factor analysis. These sources are the United States Bureau of Census, the United States Bureau of Labor Statistics, the State of North Dakota, the State of Wyoming, the State of Montana, the State of Minnesota, the State of South Dakota, and the State of Nebraska for the years 1990–2000. The data are county-level and represent the 96 counties within the study area, which were identified earlier in the chapter. Besides the five CERAtown counties, the additional 91 rural counties are needed to increase the data set to a level where factor analysis results will be valid. Thus, similar counties both in structure, i.e., population density measures, and geography are selected.

DESCRIPTION OF SELECTED SOCIO-ECONOMIC VARIABLES

Twenty-two socio-economic variables are included in the factor analysis. The selection of variables for use in this study is based on existing literature, with two significant research publications serving as the foundation. These studies identify factors influencing the implementation of information technologies, including personal computers and broadband telecommunications. These studies are: *Having All the Right Connections: Telecommunications and Rural Viability* (Korsching, et al. 2000) and *Falling Through the Net: Toward Digital Inclusion* (National Telecommunications and Information Administration 2000). Based on these studies, factors influencing the implementation can be grouped into two categories: demographic and economic.

As noted in *Falling Through the Net* (2000), the demographic variables considered important to the utilization of personal computers, Internet, and broadband telecommunications are race, education level, and age. According to the study, those who are non-white, poorly educated, and older generally do not have access to, or simply do not have the interest in using, information technologies. The leading users of Internet and computer technologies were white, college-educated persons. Economic variables considered significant indicators of information technology usage are income and economic structure. Economic structure refers to the types of businesses located within an area, such as manufacturing, service, retail, etc. According to the research of Korsching, et al. (2000), rural communities that build upon existing organizations tend to be more successful in telecommunications-based development. Thus, businesses more inclined to utilize information technologies, such as service industries, tend to reinforce use of computers and Internet by employees at home.

Although the indicators represented in the previous research studies identify demographic and limited economic variables for usage, a third set is geographic factors identified as important for economic development potential. According to Florida (2002, 12), three locational traits are important for rural communities to development within the "knowledge economy." These locational traits are:

1) Distance to an urban center: "the closer you are the better off you will be."

2) Proximity to a university: "they strengthen and bolster universities because they are talent magnets."
3) Life-style destination communities: "utilizing natural amenities to attract talented labor for development."

The variables chosen for the factor analysis reflect the demographic, economic, and locational factors designated as significant in the research. The variables are identified in Table 4-1, with raw data for each county and variable presented in Appendix A, in both geographic and spreadsheet form.

RESULTS FROM THE ANALYSIS

It is a difficult task to explain factor analysis in an uncomplicated manner. Beyond being sound methodologically, the success of this research depends on the ability of the economic development practitioner to understand and apply these tools in the field. In response to this need for academic and practical clarity, the results of the factor analysis are explained both technically and functionally.

Table 4-1. Socioeconomic Variables Important to the Implementation of Information Technologies and Broadband Telecommunications

Variable Name	*Description and Year*
pctwhite	Percentage of Population White, 2000
pctnonwh	Percentage of Population Non-White, 2000
pciprch	Percentage Change in Per Capita Income 1995-1999
pci1999	Per Capita Income, 1999
pctcol	Percent of Population with Bachelor's Degree or Higher, 1990
pctwohs	Percent of Population without High School Education, 1990
pctmanf	Percent of Employment in Manufacturing, 1999
pctsrrt	Percent of Employment in Service and Retail, 1999
rtsapc99	Retail Sales Per Capita, 1999
avgunp00	Average Unemployment Rate, 2000
pct17und	Percent of Population Age 17 and Under, 2000
pct65ovr	Percent of Population Age 65 and Over, 2000
mage2000	Median Age, 2000
tpop00	Total Population, 2000
popc9000	Percentage Change in Population, 1990-2000
psm2000	Persons Per Square Mile, 2000
intaccs	Interstate Access in County
dstnmsa	Driving Distance to Nearest Metropolitan Statistical Area
preshied	Institution of Higher Education in County
pchnfest	Percentage Change in Non-Farm Business Establishments, 1990–1999
pctbepov	Percent of Population Below Poverty Level, 1997
urban	Percent of Population Residing in the Largest City of the County, 2000

In technical terms, the use of factor analysis allows for the reduction of a larger set of initial variables to a smaller, more manageable set of derived indicators without assuming all of the variables are of equal importance (Hansen and Hemmassi 2001). The general method builds on the statistical reduction of the original indicators into their "principle components," based upon the mathematical covariance relationship among the set of original variables. After the significant variables are identified, a VARIMAX rotation, a type of orthogonal factor rotation, is conducted to produce independent factors, making the correlation between factors 0.

The analysis operates under a minimum eigenvalue of 1.0 and communality estimate of .50. After numerous factor procedures, five variables were eliminated: pchnfest, pctmanf, avgunp00, psm2000, pciprch (Appendix B). Based upon the remaining 17 variables, the factor analysis yielded four factor solutions (Table 4-2). As demonstrated in Figure 4-7, the four factors produced eigenvalues of 5.289, 4.379, 1.499, and 1.266, respectively. These four factors accounted for 73.1 percent of the total variance, indicating their importance as descriptive variables.

Table 4-2. VARIMAX Rotated Factor Matrix and Final Communalities

	Factor Pattern Loading				
	Factor 1	*Factor 2*	*Factor 3*	*Factor 4*	
Variable	*Poverty*	*Affluence*	*Regional Center*	*Accessibility*	*Communality*
pctnonwh	**0.94947**	–0.13783	–0.05168	–0.05942	.926
pct17und	**0.76761**	–0.04095	–0.18915	0.04710	.629
pctbepov	**0.69748**	**–0.51241**	–0.19280	–0.14322	.807
pct65ovr	**–0.63400**	**–0.57190**	–0.00086	–0.31094	.826
mage2000	**–0.83531**	–0.31576	–0.10454	–0.30313	.900
pctwhite	**–0.94920**	0.13609	0.05274	0.05589	.925
popc9000	0.31363	**0.80085**	0.15818	0.02596	.765
pctcol	–0.18427	**0.73846**	0.35426	0.04198	.707
pci1999	–0.23208	**0.58120**	**0.53649**	–0.19277	.717
pctwohs	0.42021	**–0.76325**	–0.22720	0.00875	.812
pctsrrt	–0.10574	0.29108	**0.78853**	–0.10171	.728
rtsapc99	–0.08607	0.24387	**0.75869**	–0.05716	.646
urban	–0.20761	0.07671	**0.68286**	0.19480	.553
tpop00	0.27098	0.38604	**0.54850**	0.47135	.745
preshied	0.28771	–0.00978	**0.52022**	0.47524	.579
intaccs	0.02954	0.13287	0.19036	**0.77096**	.649
dstnmsa	0.10318	0.16598	0.30984	**–0.62046**	.519
Variance Explained	5.289	4.379	1.499	1.266	
Cumulative Percent of Variance	31.1	56.9	65.7	73.1	

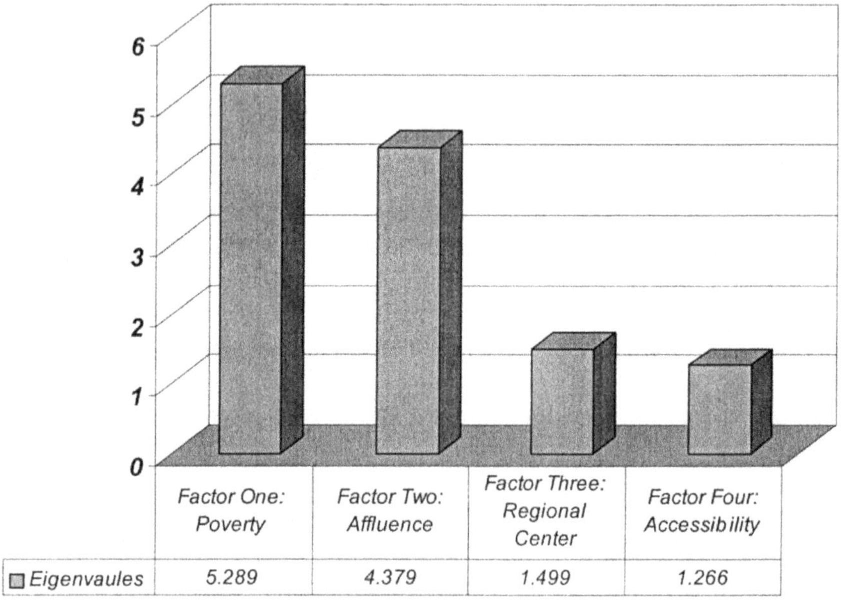

Figure 4-7. Eigenvalues for Emergent Factors.

For the economic development practitioner, factor analysis basically is a statistical tool that groups variables into a more meaningful and manageable set of factors. The VARIMAX rotation is an arithmetical way of "forcing" the entire set of variables to fit into one of the emerging factors. The use of eigenvalues and communality establishes a baseline to prove the importance of the factor or variable. For instance, factors with a value of less than one or communality of less than .50 are not significant so they are removed from consideration. Thus, based on these minimum criteria, several of the factors and variables are thrown out. It is also important to note that each time you remove a variable from consideration, it changes the values of the other factors. So the analysis must "run" several times until the final factors are identified. Variance refers to the ability of the emerging factors to describe the "overall picture" for the entire set of variables. Accordingly, a variance of 100 percent would be a statistically perfect description. The final practical explanation is this rule: the higher the factor scores the greater significance of that factor for a specific location.

Through the application of factor analysis, four significant factors emerge:

1) Factor One: Poverty
2) Factor Two: Affluence

3) Factor Three: Regional Center
4) Factor Four: Accessibility

Factor one, the most important element, accounted for 31.1 percent of the total variance (73.1 percent). As shown in Table 4-2, this factor is comprised of variables with high positive factor loadings, such as percentage of population non-white (pctnonwh), percentage of population under age 17 (pct17und), and percentage of population below poverty (pctbepov), also strong negative loadings such as percentage of population 65 and over (pct65ovr), median age of population (mage2000), and percentage of population white (pctwhite). These loadings represent the relationship between race and poverty. Thus, the first factor is labeled "Poverty." This factor clearly identifies counties that contain high Native American populations. As shown in Figure 4-8, counties with high factor scores, such as Sioux, ND (4.39), Rolette, ND (3.92), Corson, ND (3.28), and Roosevelt, MT (3.17), are home to Native American reservations. Larger negative scores belonged to counties with little racial diversity and low levels of poverty, such as Divide, ND (–1.15), McIntosh, ND (–1.07), and Nelson, ND (–1.06).

Factor two, also high in statistical importance, accounted for 25.8 percent of the total variance (73.1 percent). Factor two is representative of growth and prosperity and titled "Affluence." This factor, as noted in Table 4-2,

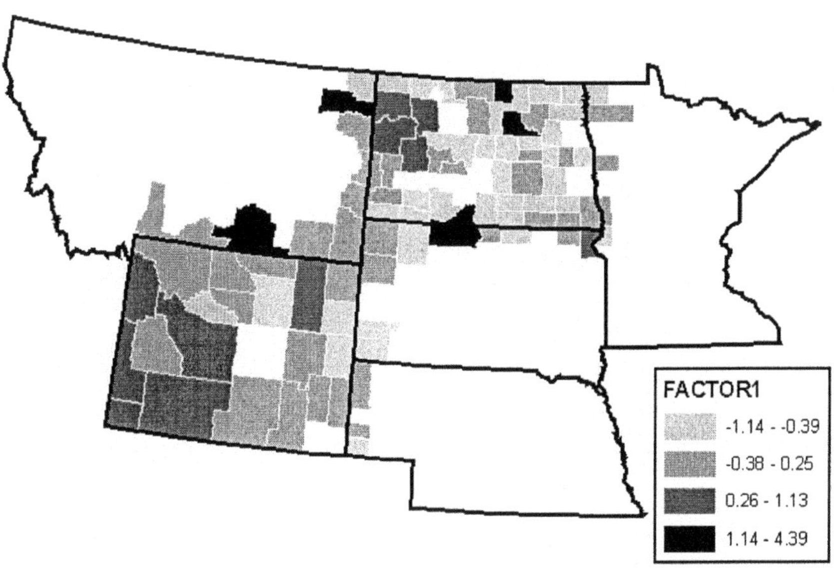

Figure 4-8. Factor Scores by County—Factor One: Poverty.

is comprised of such high factor loadings as population percentage change (popc9000), percentage of population with bachelor's degree or higher (pctcol), and per capita income (pci1999). The negative factor loadings include percentage of population without a high school education (pctwohs), percentage of population below poverty (pctbepov), and percentage of population 65 and over (pct65ovr). Counties with high factor scores represent areas of high growth, high education, and high income. As shown in Figure 4-9, counties with high positive factor scores include Teton, WY (5.27), Sublette, WY (2.77), and Albany, WY (1.80). Negative factor scores appeared in counties with declining and again populations, along with low education levels, such as Sheridan, NC (–1.90), Grant, ND (–1.56), and Emmons, ND (–1.25).

Factor three, which accounted for 8.8 percent of the total variance (73.1 percent), is comprised of such high factor loadings as percent of employment in service and retail (pctsrrt), retail sales per capita (rtsapc99), percentage of population residing in the largest town (urban), total population 2000 (tpop00), presence of an institution(s) of higher education (presihed), and per capita income (pci1999). Such variables as greater population concentration, increased retail sales per capita, high employment in service and retail, are representative of rural counties that possess many of the traits expected in urban areas. More specifically, these counties represent sub-regional trade centers and college towns. Thus, factor three is entitled "Regional Centers." As noted in Figure 4-10, counties with high factors scores are Teton, WY (3.11),

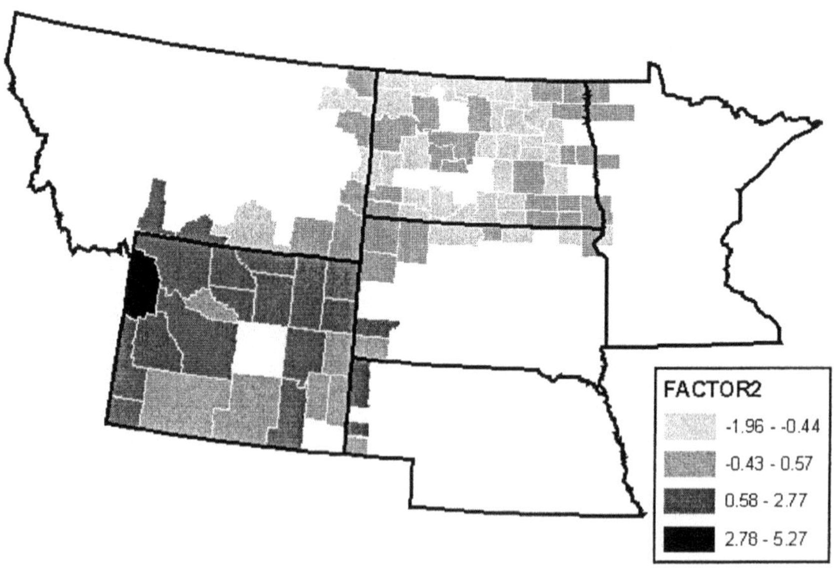

Figure 4-9. Factor Scores by County—Factor Two: Affluence.

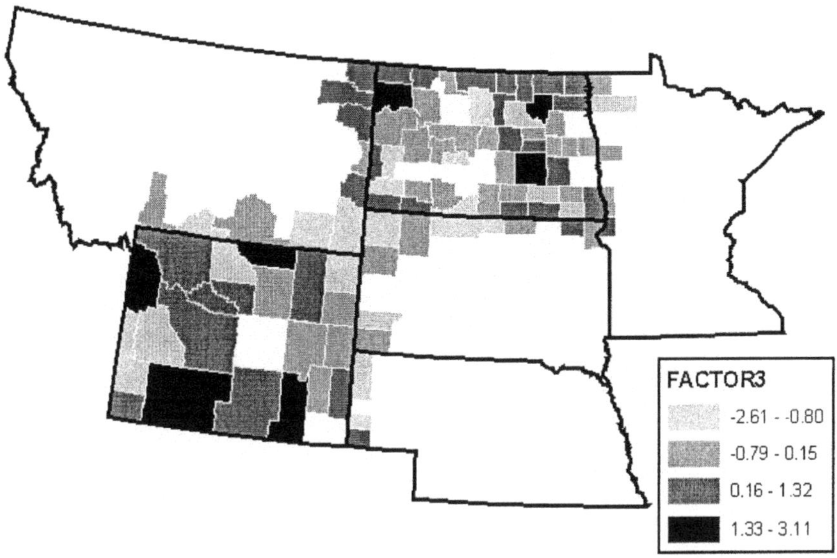

Figure 4-10. Factor Scores by County—Factor Three: Regional Center.

Williams, NC (2.51), and Stutsman, ND (2.09). It is of importance to note that factor three did not contain any significant negative factor loadings.

Factor four accounted for 7.4 percent of the total variance (73.1 percent). It is comprised of two factor loadings, interstate accessibility within the county (intacss), positively related, and driving distance to nearest MSA (dstnmsa), negatively related. This factor is representative of counties located along interstate highways and near metropolitan areas. Due to the geographic importance embodied in factor four, it is labeled "accessibility." As exemplified in Figure 4-11, counties with high factor scoring include peripheral counties such as Albany, WY (2.67), Campbell, WY (2.16), and Stutsman, NC (2.09).

THE "BROADBAND DEVELOPMENT INDEX"

The analysis produces four distinct factors with varying degrees of importance: poverty, affluence, regional center, and accessibility. When relating the emergent factors to the existing literature on broadband and information technology implementation, three of the factors are positively related to use, affluence, regional center, and location, while poverty, the most significant factor, is negatively related.

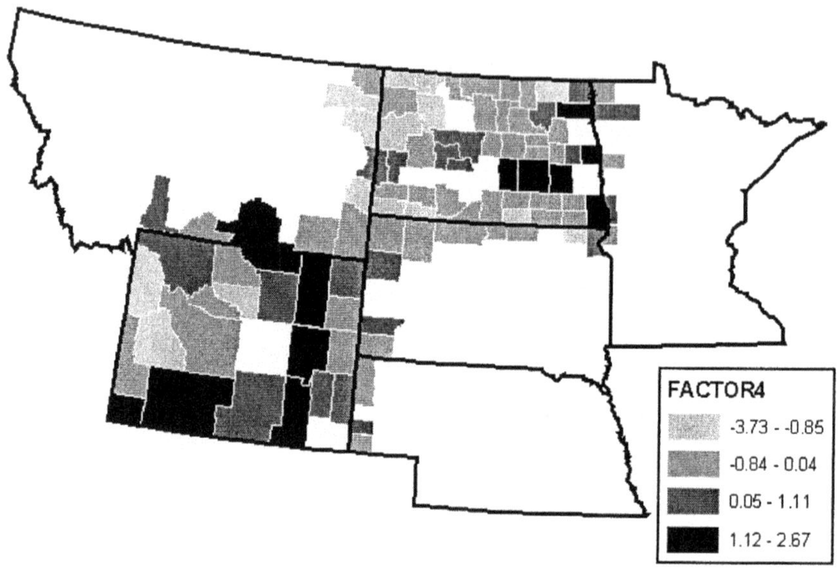

Figure 4-11. Factor Scores by County—Factor Four: Accessibility.

Once the relevant factors are identified, it is possible to develop an index that identifies the potential of study area counties to utilize information technologies for economic development success. The resulting index can be used for regional economic development planning to identify rural communities with technology-based development possibilities. Therefore, the "broadband development index" is offered as a practical development tool.

The "broadband development index" (BDI) is a weighted composite indicator used to determine the differences or degrees of development potential based upon the use of information technologies and broadband telecommunications. It is expressed mathematically as:

$$\text{BDI Score} = \Sigma(Xi)(fi)$$

where: Xi represents the ith factor score and fi percentage of the total variance in the ith factor explains (Hansen and Hemmasi 2001; Weng 1997). In simplified terms, the index is created by the sum of the factor scores weighted by their individual variance. For example, factor one, poverty, would represent a higher proportion of the index score than factor four, accessibility.

Due to the negative relationship between broadband use and poverty (factor one), the adjusted factor scores for poverty are multiplied by (–1). This step is proper because the condition of poverty, through the original BDI formula, would be accepted as a positive factor, producing an inaccurate representation of poverty as a contributor to development. Due to this adjustment, a higher positive score means the greater the development potential based upon the emergent factors. The broadband development index scores are presented for each of the study area counties by location in Figure 4-12 and score in Figure 4-13. A detailed table of the index and adjusted factor scores are contained in Appendix C.

According to the index scores, the five counties with the greatest development potential based upon technology implementation are Teton, WY (1.131), Albany, WY (0.730), Park, MT (0.475), Sheridan, WY (0.466), and Sublette, WY (0.441). Geographically speaking, these counties possess great natural amenities, along with high per capita incomes, high education levels, and educational institutions. The highest scoring county in North Dakota is Traill, ND (0.332), which is home to one of the participating CERAtown communities.

Counties with the lowest index scores are: Sioux, ND (–1.404), Rolette, ND (–1.229), Corson, SD (–1.139), Roosevelt, MT (–1.030), and Benson, ND (= 0.940). These counties have high Native American populations,

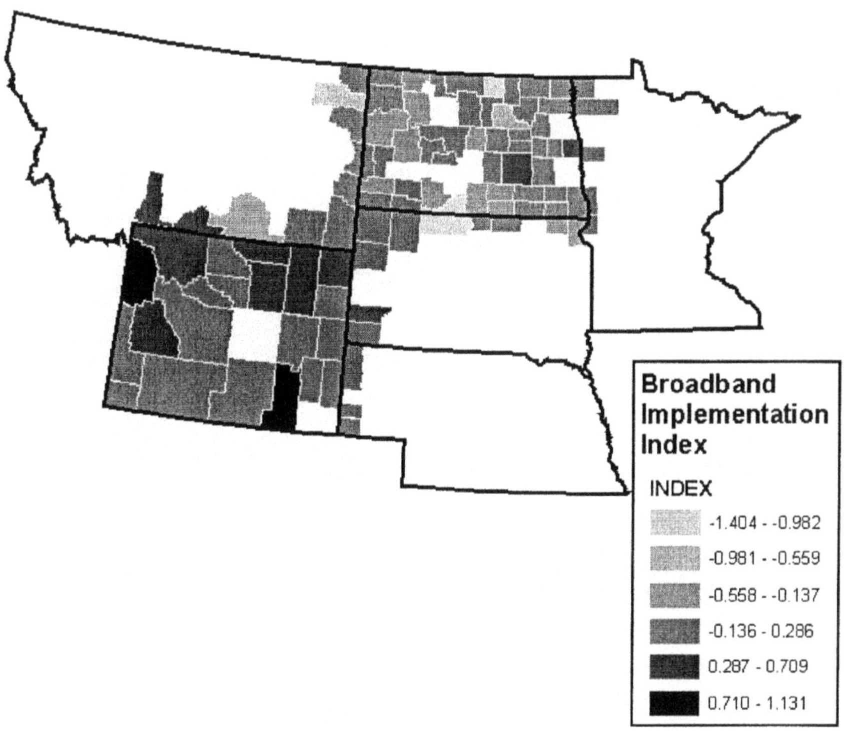

Figure 4-12. Broadband Development Index by Location.

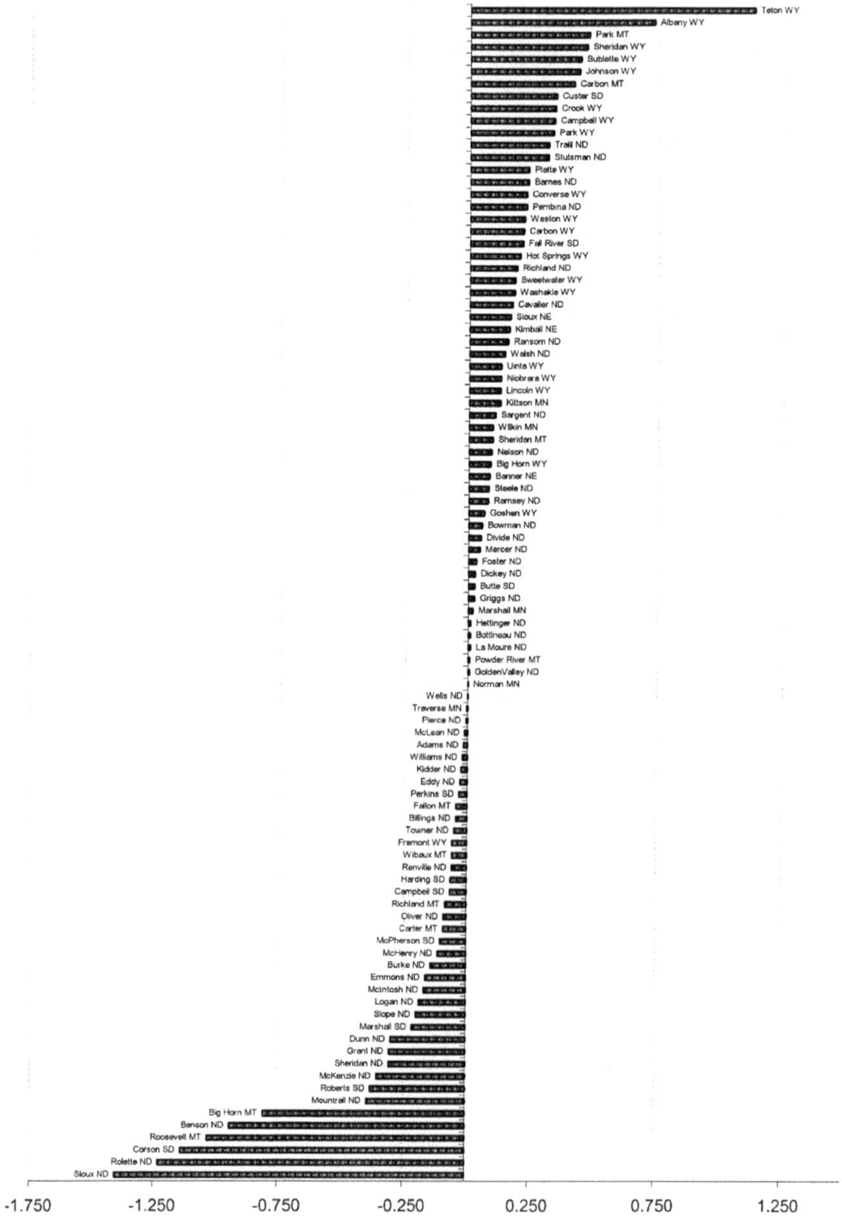

Figure 4-13. Broadband Development Index Scores by County.

high poverty levels, and lower levels of educational attainment. The lowest score for a participating CERAtown community is McKenzie County, ND (–0.354).

The BDI allows for the ranking of counties that contain communities participating in the CERA project. According to the index scores, the CERAtown counties rank in the following order: Park, WY (0.337), Traill, ND (0.332), Converse, WY (0.225), Niobrara, WY (0.126), and McKenzie, ND (–0.354). Since McKenzie County, ND is the only location with a negative score, it is identified as the CERAtown community with the lowest broadband-based development potential. On the other hand, the most potentially successful location for telecommunication-based economic development is a close competition between the CERA communities of Park County, Wyoming and Traill County, North Dakota.

In order to optimize the results of the qualitative research, it is important to observe communities that differ in the economic, demographic, and geographic factors identified in the analysis. Yet, it is important to keep certain traits, especially population, similar in order to draw a fair comparison. For example, it is not rational to compare Powell, Wyoming, Park County, to Watford City, North Dakota, McKenzie County, because Powell has a population that is five times greater than Watford City. Due to similar populations and pronounced differences in geography and economic structure, the second ranked CERA community of Traill County, North Dakota, is used for the qualitative analysis.

CONCLUSION

This chapter produces three important tools for the development of a model for rural economic development planning based upon broadband telecommunication. First, a theoretical framework for constructing a rural economic development planning model, based upon the ideas of Porter's theory of competitive advantage and The University of Minnesota's "rural knowledge cluster" model. Second, the establishment of the CERA program as the focus of both quantitative and qualitative research. The final tool produced is the development of a broadband index that ranks the factors that contribute to broadband technology use, as well as establishing a foundation for justifying the selection of the appropriate CERAtown communities for further research.

Through the review of literature and results from the factor analysis, it is evident that certain economic, demographic, and geographic factors are important in the implementation of information technologies and broadband telecommunications in rural areas. In addition to implementation, another influential factor is the presence of institutions that retain and attract "talented" labor, such as colleges, universities, and research institutions.

Though the results of the quantitative analysis generate a starting point for the construction of a rural economic development planning model, it does not provide a complete representation of cluster-based economic development. Issues such as local leadership, entrepreneurship, and other non-quantifiable factors are significant and must be considered.

In Chapter Five, the CERAtown communities of Mayville, North Dakota and Watford City, North Dakota, along with Ennis, Ireland and LaGrange, Georgia, serve as case study location where qualitative research completes the construction of a model for rural economic development planning based upon broadband telecommunications.

Chapter Five

Local Leadership and Policy in Broadband Telecommunications-based Rural Economic Development Planning

Sports prognosticators analyze the attributes of football teams to predict the outcomes of games. These areas of analysis include offense, defense, special teams, and so on. Though these categories have statistical foundations for comparison, the category that often influences the outcome is "intangibles." These "intangibles" are the non-statistical factors that play a significant role in the result of the game. As in football, "intangibles" play an important role on the "playing field" of economic development. Of these "intangibles," one of the most essential within the practice of economic development is leadership.

The literature recognizes the significance of leadership in local economic development. According to Thornburgh (1998, 298), "without creative, sustained local leadership, federal help alone can have little enduring impact on local economic conditions." Also, the research studies identify local leadership as a critical factor in improving the economic conditions of a community (Raagmaa 2001; Aristigueta and Frenandez 1998; Winter 1996). In addition to changing economic conditions, leadership is a force that determines the success of local economic development. Malcolm Williams notes:

> Within any given region, communities both large and small are essentially the same. They have the same environment, natural resources, market access and demographics. The one factor that makes them different is leadership. Why do similar towns with similar characteristics have different results in their economic development efforts? The answer is leadership. All towns say that "we have good people, good workers, good leadership! (Williams 1990, 4)

Due to the importance of leadership, this chapter studies the potential role of local leadership in the application of broadband telecommunications for rural economic development.

Through the application of the cluster framework, i.e. Porter's Diamond Model (Porter 1990), the identification of "intangibles" rests within the two supplemental categories of government/policy and chance. This research section addresses the idea of chance through the supposition that local leadership and policy are key "intangibles" in the success or failure of rural economic development initiatives based on broadband telecommunications. In addition to local economic development strategies addressed in this chapter, Chapter Three acknowledges the importance and content of federal and state policies for broadband telecommunications-based economic development. Due to the value of leadership for local economic development, the following research question is proposed: What is the importance of local leadership and policy in the creation of a rural economic development model based on broadband telecommunications?

To properly attend to the research question, the chapter is divided into the following sections:

1) Research Design and Methodology
2) Exploring the Role of Local Leadership and Strategies in Broadband Telecommunications-Based Rural Economic Development Projects: A Multiple Case Study Approach

    a. Ennis, Ireland
    b. LaGrange, Georgia
    c. Watford City, North Dakota
    d. Mayville, North Dakota

In addition to addressing the proposed research question, this chapter includes the culmination of the entire research project: a structural model for utilizing broadband telecommunications for rural economic development. Therefore, the concluding segment of this chapter is the following: Utilizing Broadband Telecommunications for Economic Development: A Structural Model. The model includes policy recommendations, community evaluations based on cluster analysis, and additional planning strategies revealed through the qualitative research of this chapter.

RESEARCH DESIGN AND METHODOLOGY

Leadership is not a readily identifiable feature within localities; more specifically, it is not a quantifiable characteristic. In order to research the importance of leadership in rural communities, it is necessary to construct a research design that enables personal contact with local citizens to extract their opinions about the value of leadership. Based on the "intangible" nature of leadership, the appropriate research method is a qualitative design.

Qualitative research is broadly defined as, "any kind of research that produces findings not arrived at by means of statistical procedures or other means of quantification" (Strauss and Corbin 1990, 17). According to Hoepfl (1997), a qualitative research design possesses three strengths:

1) Depth and detail—may not get as much depth in a standardized questionnaire.
2) Openness—can generate new theories and recognize phenomena ignored by most or all previous researchers and literature.
3) Helps people see the worldview of those studies—simulating their experience of the world.

Since leadership is a difficult variable to ascertain, qualitative research methods offer the most appropriate structure to identify the importance of leadership within communities utilizing broadband telecommunications for job creation.

Of the numerous qualitative research methods available, this study utilizes a multiple case study design. Multiple case study design employs "replication logic," where identical survey instruments are used in several locations (Yin 1994, 36). By structuring the case studies around a single survey instrument, evidence from the research is considered more compelling and the overall study is considered more robust (Firestone 1987). Also, the multiple case technique is a commonly accepted approach for the development of policy-oriented theory (Yin 1994).

Based on the multiple case approach, this research follows these steps:

1) Step One: Design the survey instrument and develop questions that obtain the needed information.
2) Step Two: Identify both the case study locations and appropriate persons to interview.
3) Step Three: Conduct interviews using the single survey instrument.
4) Step Four: Interpret results.

The use of the multiple case study approach to identify the importance of leadership requires a sound survey instrument. In response, this study uses in-person, tape-recorded interviews to collect the needed responses. Through careful consideration and construction, the "leadership survey" and interview script acquires the necessary information from the interviewees about leadership and broadband-based economic development. The "leadership survey" consists of 15 questions, which are divided into two sections. The first section is a series of six questions requiring a designated response, followed by a sequence of open-ended questions. This set of questions focuses on two issues, leadership and perceived success of the broadband economic development project. The second section consists of nine open-ended questions. These questions extract information about innovative strategies and personal applications of available information technologies, e.g. broadband Internet, personal computers, etc. Appendix D contains a copy of the "leadership survey" and interview script.

The next progression in the research structure is the identification of case study locations and interview participants. The case study locations for qualitative research are the same introduced in the preceding chapters. The case study communities are Ennis, Ireland; LaGrange, Georgia; Watford City, North Dakota; and Mayville, North Dakota. Once again, these communities are recognized for their application of broadband telecommunications for rural economic development. Also, selecting an international example increases the validity of the results and proves itself important in the creation of a model designed to be universally applicable in rural communities.

Following the "replication logic" of the multiple case design, a designated set of representatives are interviewed within each of the case study communities. These representatives include one person from each of the following sectors: small-business representative; agricultural or service sector representative; project representative; education representative; and economic development representative. The representatives are selected because each sector is well positioned to identify local leaders and in particular, how leadership, or lack thereof, affects the broadband telecommunications development strategies within each of the case study communities. A total of 20 people participated in the interview; five persons per case community, one representative from each sector. Figure 5-1 displays the research design.

EXPLORING THE ROLE OF LEADERSHIP IN BROADBAND TELECOMMUNICATIONS-BASED RURAL ECONOMIC DEVELOPMENT PROJECTS: A MULTIPLE CASE STUDY APPROACH

This section presents results from the interview responses and fieldwork within the case studies communities of Ennis, Ireland; LaGrange, Georgia; Watford City, North Dakota; and Mayville, North Dakota. These case studies represent a variety of approaches to technology-based economic development

Figure 5-1. Research Design.

initiatives, from the highly funded Ennis, Ireland project to the non-funded CERAtown communities of Watford City and Mayville, North Dakota.

The case studies are presented by location and divided into the following sections:

1) Introduction
2) Survey Results
3) Emerging Themes

The introduction presents both an overview of the community and specifics about the broadband-based economic development project. The survey results include interview responses from each representative sector.

Ennis, Ireland

Ennis, County Clare, Ireland is a town of approximately 18,000 persons, situated in Midwestern region of the country (Figure 5-2). The city of Ennis is located within the Shannon Development Region, a noted entity within the Irish economic development community. Ennis functions as the regional trade center for the surrounding hinterland with the service industry accounting for nearly half of its employment base (McQuillian 2001). According to Waters and Byrne (2000), Ennis possesses a bright future with its high population of well-educated, young people.

In 1996, *eircom*, formerly Telecom Eirann, launched a national competition designed to accelerate Ireland's movement into the information economy. The project established two goals: 1) "to saturate a town with 21st Century telecommunications technology to see how people come to terms with and use the technology," and 2) "encourage the town to trial new technologies and applications" (*eircom* 1996, 6). The project provided a 19 million (US$18.5 million) investment over five years and awarded the winning community the title of "Ireland's Information Age Town." Of the 46 town entries, Ennis was named the winner of the contest in September of 1997.

On winning the contest, *eircom* directed the Ennis community to answer five key questions:

1) What happens when every home has advanced telecommunication services?
2) What happens when every business, large and small, has broadband access to the Internet?

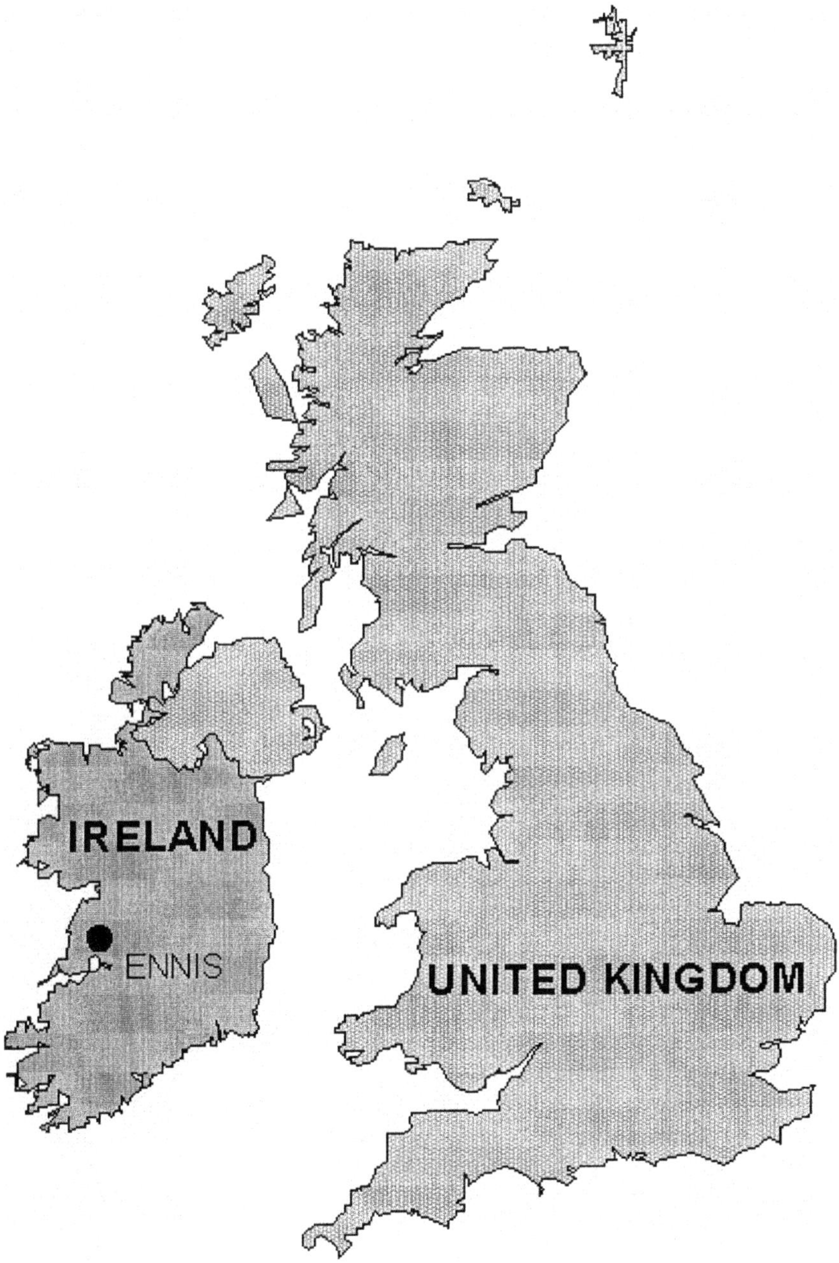

Figure 5-2. Geographic Location of Ennis, Ireland.

3) What happens when every student in the education system, from age five and beyond, has intensive access to computer and advanced Internet connections?
4) What happens when public services, from libraries to healthcare facilities, are fully equipped to exploit the potential of information technologies?
5) What happens when the majority of all households have a personal computer and Internet connection?

Thanks to the media attention received through the competition, Ennis became known as "the largest community technology project in the world" (Industry Ennis 2001, 6).

But the potential of Ennis the "Information Age Town," stretches far beyond the benefits of increased media coverage. According to Erkki Liikanen, European Union (EU) Commissioner of Enterprise and Information, Ennis "is designed to play a central role in the agenda of economic and social renewal in Europe and is a key element in the strategy of the Commission to modernize the European economy" (Industry Ennis 2001, 8). Due to its stature and structure, the "Information Age Town" of Ennis is a valid selection for study.

The Ennis interviews provide great insight into the role of leadership before and after the awarding of the "information age" project. Without exception, the representatives observe local leadership as vital in the success of Ennis. These results are presented in Table 5-1.

The economic development representative identified himself/herself as directly involved in the project and viewed the project successful. The representative noted the project as "successful" because of the attainment of advanced technological infrastructure, large amount of direct funding, and the creation of jobs and wealth in Ennis. Based on the interview transcript, the leadership forces are divided into pre and post information age project designations. According to the economic development representative:

> Local citizens, myself included, worked around the clock during the proposal period. It was not elected officials driving the process as such. It opened up a channel where a new type of leadership emerged as such, where tentacles reached to every corner of the community through grassroots efforts. It was through the proposal process, the community developed a sense of ownership in the project. This leadership exemplifies the support of the community for the project.

Yet, after the awarding of the project, the economic development representative felt that leadership transferred to the staff hired to direct the Ennis project.

Table 5-1. Answers to Designated Response Questions—Ennis, Ireland

Survey Questions	*Economic Development Representative*	*Education Representative*	*Small Business Representative*	*Agricultural/ Service Sector Representative*	*Project Representative*
Are you directly involved in the IT/broadband project?	**YES**	**NO**	**NO**	**NO**	**YES**
Do you view the project successful?	**YES**	**YES**	**YES**	**YES**	**YES**
Does the community support the project?	**YES**	**YES**	**YES**	**YES**	**YES**
Did the project increase the economic development activity in the community?	**YES**	**YES**	**YES**	**YES**	**YES**
In your opinion, has the project been driven from the bottom up or top down?	**TOP-DOWN**	**TOP-DOWN**	**BOTH**	**TOP-DOWN**	**BOTH**
Do you credit local leadership as the most critical factor in the success or failure of the project?	**YES**	**YES**	**YES**	**YES**	**YES**

The project representative interview resulted in similar responses, but it was felt that the need for leadership diminished with the introduction of information technologies. The project representative said:

> The initial work on the project needed a vast network of local leaders. Luckily, this leadership exists in Ennis. So initially, top-down leadership among community members meant a great deal to win the award. Yet, it seemed that as the technology arrived, people began focusing on the technology itself, learning and developing it, somewhat in a bottom-up fashion.

The project representative said a conscious decision was made in the initial stages of the entire project to merge the bottom-up and top-down approaches for community development. According to the representative, "we knew an initial top-down approach was essential to get things started, but the strategies implemented within the project, such as computers in each home, had to be bottom-up."

Those not directly involved in the project, the small business representative, service/agricultural representative, and education representative, viewed the project successful and credited local leadership as an essential factor. The education representative felt that students benefited dramatically from the influx of technology and the project raised the potential of the community to retain the students through increasing job opportunities. The small business representative supported the idea:

> Thanks to the project and its technology, a lot of younger business people have been given the confidence to expand their businesses in ways that just would not have been possible before. We have new facilities and businesses in town that are worth a hell of a lot. They would not be here if not for the improved infrastructure. It (the infrastructure from the project) has been good for virtually every business and attracts businesses from outside the area.

The agricultural representative recognized local leadership as key to involving all sectors of the economy. Based on the responses, the agricultural representative did not intend to use computer technology in any form until the Ennis project hosted a party for the farmers in the area. The agricultural representative said:

> I decided that if my neighbors (farmers) would give it a try, so would I. Triona (an "Information Age" employee) made me feel like I could do it and if I couldn't, she would send out someone to help me. The help was nice. It was usually a local student, and more often than not someone I knew. Without local help, I would have given up half way through setting up the computer.

Interviews with individuals not directly involved in the project reflect the importance of the top-down leadership on the spread of the "grassroots" acceptance of the information technologies.

Overall, the Ennis case study research produces three important developments. First, leadership that encourages community involvement is necessary for success. It is best described as a top-down leadership strategy, where community members feel like it is bottom-up or neighborly. Second, there must be local technical assistance. Based on the responses, dialing a toll free number for help would have prevented the use of the computer and Internet. This is particularly important for rural citizens "taking a chance" on information technologies. Third, funding does make a difference. The comprehensive nature of the project eliminates the economic barriers to technology usage, such as computer ownership and Internet access. Ennis, through its direct investment, is an "idealized" representation of technology-based economic development. As exemplified with the agricultural representative, no matter how much money you invest in a community, it is clear that local leadership can make technology-based development initiatives succeed or fail.

LaGrange, Georgia

LaGrange, Georgia, the seat of Troup County, is located 60 miles southwest of Atlanta with a population of approximately 27,000 residents. Its economic history is that of a typical southern textile community, yet its economic development success is far greater than most southern textile towns. LaGrange is an entrepreneurial city with no local property taxes and home to more Fortune 500 companies per capita than any other city its size (City of LaGrange 1999). Due to its geographic advantages and entrepreneurial spirit, LaGrange continues to grow both in population and industry. Figure 5-3 presents the geographic location of LaGrange, Georgia.

In the early 1990's, the City of LaGrange recognized the need for advanced telecommunications infrastructure. According to a Harvard Business School Study (2001, 2):

> They knew their ability to keep existing industry in LaGrange as well as the ability to attract future industry depended heavily on the capability to deliver voice and data communications in increasing quantities and increasing speeds.

In response, the city formed a "Local Telecommunications Committee" to study the capabilities and components of the local telecommunications infrastructure (City of LaGrange 1999). The report found that the city lacked a local point of presence (POP), little fiber in the network, and the LaGrange

Figure 5-3. Geographic Location of LaGrange, Georgia.

switching point was analog, not digital. Based on the findings of the study, the city began searching for a private telecommunications partner to upgrade the telecommunications infrastructure. Through an agreement with ITC-Deltacom, the company provided local digital switching capacity and the City of LaGrange made fiber optic lines available to businesses.

Thanks to the agreement with ITC-Deltacom, the City of LaGrange began to profit by selling excess bandwidth to local commercial customers. In 1994, the first contract with a local business resulted in over $45,000 in gross revenue. According to Harvard Business School (2001), over the next six years, LaGrange developed a fiber ring to serve over 60 commercial customers and the resulting gross revenues were over $1 million a year. Consequently, LaGrange solved its infrastructure problems, profited from the arrangement, and retained, expanded, and attracted businesses.

Though the broadband capabilities were available to large businesses, most citizens and small business did not have access. In an effort to bridge this gap, the City of LaGrange entered into an agreement with Charter communications to offer high speed Internet access through television cable, at no cost to citizens. On top of the free access, Charter and LaGrange rolled out a new service called "WorldGate Internet." The WorldGate service provides a wireless keyboard to every household that interfaces with an Internet-enabling device hooked to the television set. The costs associated with WorldGate, approximately $400,000 a year, are covered by the City of LaGrange. Thus, making the service and hardware free to all residents.

These innovative and progressive steps earned LaGrange, Georgia international acclaim as a local economic development success story. The City of LaGrange won the World Teleport Association's Intelligent Community of the Year Award in 2001 for its deployment of free high-speed Internet service to every home and business in the community. LaGrange is also recognized in economic development literature as one of the most progressive and successful rural cities in the use of telecommunications and information technologies for economic development (Youtie 2000). Based on its reputation of success, LaGrange is a convincing location to study the importance of local leadership and policy strategies.

The LaGrange interviews solidify the role of leadership in progressing the economic development of the city. All of the participants agreed that leadership contributed to the success of the broadband initiative in LaGrange. Table 5-2 presents answers to the designated response questions.

The economic development representative of LaGrange credited "visionary leadership" and "timing" as key to the city's success. According to the representative:

> Without visionary leadership, particularly in the early 1990s, our economic situation would probably be similar to other textile cities. It convinced local industries we were concerned. And we appeared progressive to companies looking for a new location. We also took advantage of the market conditions within the telecommunications industry to cut a great deal. You could not do the same thing in today's market.

Likewise, both the project representative and agricultural/service representative acknowledged the importance of local leadership in the original deal with ITC-Deltacom. The education representative responded to the question of leadership with the following comment:

> As far as the school system goes, we have benefited greatly from the enhanced infrastructure within the city. I don't know who got that first deal done to get us the fiber, but they deserve a lot of credit. Seems like our success snowballed from that deal.

The small business representative said:

> Thanks to our leadership, we were ahead of the curve in telecommunications and Internet. I even heard we had better stuff than bigger cities around us, like Columbus. From the initial fiber up to this new cable Internet system, it's our leaders that made it happen. If we had left it up to the telephone companies, it would have never happened as fast as it did. We would be years behind instead of ahead.

In short, leadership proved the most important factor for the success of the initial project and the WorldGate free Internet system.

The interviews also reveal leadership coming from the top-down, not the bottom-up. Those interviewed representatives found it almost impossible to envision leadership coming from below in LaGrange. Even with the availability of free high-speed Internet and WorldGate system, everyone does not use it. According to the project representative, approximately 50% of the households utilize the service. The education representative felt that a "fear of big brother" mentality exists among many of the uneducated citizens when it comes to the Internet. The representative said:

> From what I have heard, many of the poor and uneducated citizens are afraid to use the system because they think the city can see everything they are doing in the home and what they are looking at on the Internet.

The small business representative said, "LaGrange, although it has some urban problems, thinks of itself as rural. I would say our residents wait for things to happen."

With regard to innovative strategies, the interviews do not produce any major results. Yet, through the comments and attitudes of the representatives an interesting strategy is revealed. At numerous times in the interviews, credit is given to the economic development office for "seeing the big picture." In a community whose success is directly credited to "visionary" leadership, the economic development office provided the needed expertise and foresight to move LaGrange into the future. Thus, using the interview responses as an indicator, without this top-down push of leadership, LaGrange would be just

Table 5-2. Answers to Designated Response Questions—LaGrange, Georgia

Survey Questions	*Economic Development Representative*	*Education Representative*	*Small Business Representative*	*Agricultural/ Service Sector Representative*	*Project Representative*
Are you directly involved in the IT/broadband project?	**YES**	**NO**	**NO**	**YES**	**YES**
Do you view the project successful?	**YES**	**YES**	**YES**	**YES**	**YES**
Does the community support the project?	**YES**	**YES**	**YES**	**YES**	**YES**
Did the project increase the economic development activity in the community?	**YES**	**YES**	**YES**	**YES**	**YES**
In your opinion, has the project been driven from the bottom up or top down?	**TOP-DOWN**	**TOP-DOWN**	**TOP-DOWN**	**TOP-DOWN**	**TOP-DOWN**
Do you credit local leadership as the most critical factor in the success or failure of the project?	**YES**	**YES**	**YES**	**YES**	**YES**

another textile dependent southern city, instead of a world-renowned example of telecommunications-based economic development.

Overall, the LaGrange case study reveals the power of "visionary" leadership and how it can influence the future of a community. It also reinforces the importance of an economic development official in recognizing and implementing strategies that position the city for growth. In short, the City of LaGrange had the right leadership at the right time. The economic development representative said it best, "everything lined up for us at the perfect time and as you know, a little luck never hurts."

Watford City, North Dakota

As noted in Chapter Four, Watford City, North Dakota is a ranching and oil community located in McKenzie County, just north of Theodore Roosevelt National Park in western North Dakota. The population of Watford City is 1,435 (U.S. Census 2000). Watford City is one of the original participants of the CERAtown program. Before the CERAtown project, Watford City developed a "Community Technology Plan" which acknowledged the difficulties of attracting high-tech industries to rural locations with inadequate telecommunications infrastructure. Its role in the CERA project is to create a community-based technology and education/training center, utilizing the high-speed telecommunications capabilities of the local high school. Figure 5-4 presents the geographic location of Watford City, North Dakota.

Figure 5-4. Geographic Location of Watford City, North Dakota.

Based on the results of the "broadband development index" in the preceding chapter, Watford City is the lowest ranking of the CERAtown communities. This is due to its lack of institutions and poor geographic location. Yet, the index score does not tell the entire story of Watford City. In fact, the opposite is taking place in Watford City. It is recognized as the leading CERAtown community because of its ability to attract and encourage business development through the use of broadband telecommunications and information technologies. In recognition of its success, Watford City won the North Dakota City of the Year Award from the North Dakota League of Cities in 1999.

Watford City attracted four new employers to town through its ability to meet the advanced telecommunications needs of service businesses. These employers include Midland Atlas, Cross Consulting Group, Impact, and First International Bank and Trust.

Through the efforts of the local economic development organizations Midland Atlas relocated to Watford City from Milbank, South Dakota. Midland Atlas publishes county maps, township by township, with in-depth information on current land ownership. The company markets the maps to real-estate agents throughout the Midwestern United States. The company currently employs 18 people.

Cross Consulting Group is a strategic information technology and management-consulting firm expanding its operation from Sebeka, Minnesota. According to Nick Debronsky, President and Chief Information Officer of Cross:

> Watford City's infrastructure and trainable workforce were important for choosing this location. The rural setting allows them to provide services at a lower cost than metropolitan areas, and the prospect of high paying employment is expected to retain residents as well as attract newcomers to the area (McKenzie County Job Development Authority 2001, 3).

Cross Consulting provides services to the State of North Dakota Information Technology Division, such as programming and technical support for all state agencies. According to Jo Marie Sellner, the Watford City project leader, Cross will hire 20 employees in 2001 and eventually move to 30–50 programmers at full employment.

Impact of Watford City is a telemarketing company that employs 25 people. Impact provides sales support to numerous Fortune 500 companies and national organizations, including the American Bar Association. The business came to Watford City in 1998 and according to Mark Olson, Owner of Impact, "The Watford City center is one of our highest sales per hour. I sometimes have clients request the Watford City location specifically because of their professionalism and results" (McKenzie County Job Development Authority 2001, 3).

First International Bank and Trust in Watford City uses technology to serve its customers. Watford City serves as corporate headquarters for the bank, with branch offices in Fargo, West Fargo, Minot, Williston, Harvey, Fessenden, Killdeer, Alexander, and Bowden, North Dakota, as well as Scottsdale and Mesa, Arizona. First International boasts $25 to $30 million in annual volume and employs over 200 people. Electronic commerce allows corporate headquarters to remain in Watford City. The bank uses an ISDN (Interactive Satellite Digital Network) connected through a T1 connection for transactions and meetings. In Watford City, First International employs 40 people.

On the surface, as demonstrated by the index score, Watford City does not have the recognized traits for economic development success within the information economy. Yet, this rural city created over 100 new jobs in three years. It is evident Watford City is doing something right. It is the purpose of this case study to determine if leadership and local strategies are the "intangibles" driving the success of Watford City.

Based on the interview responses contained in Table 5-3, Watford City incorporates leadership, technological know-how, and economic development planning to create jobs in the community. The first important contribution to the transformation of Watford City is leadership. According to the agricultural/service representative:

> The economic development office worked extremely hard to bring us into the community, but more importantly, they continue to care about how we are doing . . . Their hard work transformed this city.

The project and economic development representative cited state and local leadership as key factors in community success. Though the CERA project did not receive the anticipated funding from the federal government, it succeeded in stirring up the attention of state leaders. In spite of the lack of federal support, Watford City carried on with the CERA plan by using local funds to complete the broadband installation project. The governmental bodies of Watford City and Traill County partnered to install a state of the art telecommunications infrastructure in the city focusing on high-speed Internet service to educational facilities and businesses. According to the education representative:

> The infrastructure became a source of pride for the community. Though some people were against spending the money, once their children began benefiting from the technology, resistance disappeared. The most important leadership action, in my mind, came from spending the money.

Table 5-3. Answers to Designated Response Questions—Watford City, North Dakota

Survey Questions	*Economic Development Representative*	*Education Representative*	*Small Business Representative*	*Agricultural/ Service Sector Representative*	*Project Representative*
Are you directly involved in the IT/broadband project?	**YES**	**NO**	**NO**	**NO**	**YES**
Do you view the project successful?	**YES**	**YES**	**YES**	**YES**	**YES**
Does the community support the project?	**YES**	**YES**	**YES**	**YES**	**YES**
Did the project increase the economic development activity in the community?	**YES**	**YES**	**YES**	**YES**	**YES**
In your opinion, has the project been driven from the bottom up or top down?	**TOP-DOWN**	**BOTTOM-UP**	**BOTTOM-UP**	**TOP-DOWN**	**TOP-DOWN**
Do you credit local leadership as the most critical factor in the success or failure of the project?	**YES**	**YES**	**YES**	**YES**	**YES**

The economic development representative reiterated the importance of the investment. The economic development representative said, "Without local leadership nothing would have happened, without the full support of all partners (governmental), especially in terms of funding, it would have died before it ever started."

In terms of leadership structure, opinions are mixed about top-down and bottom-up approaches. The economic development, project, and agricultural/service representatives view leadership as coming from the top-down. The project representative said:

> People were not aware of the technological capabilities. They needed someone to get the ball rolling . . . The capabilities of information technology are not readily understood in most rural places, Watford City was no exception.

In contrast, the education and small business representatives felt the leadership push was from the "bottom-up." The small business representative said:

> I knew updated telecommunications would mean a great deal to my business. I tried to get Internet service in the mid-90s and couldn't. Many businesses wanted it and needed it. So I would say much of the leadership push came from the bottom.

The education representative said, "Our pioneer spirit guided the dream of making our little city the technology center of prairie country."

The interviews bring to light innovative strategies contributing to the success of Watford City. First, Watford City partnered with the local high school to offer training classes for citizens and even went as far as giving residents the opportunity to earn various Microsoft certifications. The local education system also serves as the "go to" place for questions and assistance with computer and Internet technologies. Second, the economic development community abandoned the idea of recruiting new manufacturing businesses. The economic development leaders recognized that Watford City was non-competitive because of its geographic location. In response, Watford City implemented an economic development strategy based on improving the "quality of life" for residents. This meant focusing efforts and funding on such projects as the construction of a local movie theater and steak house.

Overall, Watford City provides meaningful results for this research. It is evident that cooperative leadership contributes greatly to the success of Watford City. The interview responses suggest dual forces of top-down and bottom-up approaches were taking place at the same time. This appears to have positive effects on the evolution of Watford City from 1998 to present.

In addition to leadership, partnering with local educational institutions proves beneficial, not only for training, but for building a "safety net" for rural citizens new to information technologies. In the end, an economic development strategy focusing on "quality of life" issues resulted in positive results for Watford City.

Mayville, North Dakota

Mayville, North Dakota, in Traill County, is located in the Red River Valley of eastern North Dakota and boasts a population of 1,953 (U.S. Census 2000). It is a farming community and home to North Dakota's smallest institution of higher education, Mayville State University. Based on its high "broadband development index" score, Mayville offers many of the locational and institutional advantages for implementing a telecommunications-based economic development strategy. Perhaps its greatest advantage lies in its geographic location. Mayville has access to Interstate Highway 29 and is within 55 miles of North Dakota's two largest cities, Fargo and Grand Forks. Figure 5-5 presents the geographic location of Mayville.

Like Watford City, Mayville is one of the original participants of the CERA program. The CERA trial project for the community is the establishment of a business incubator, housed at Mayville State University, to support

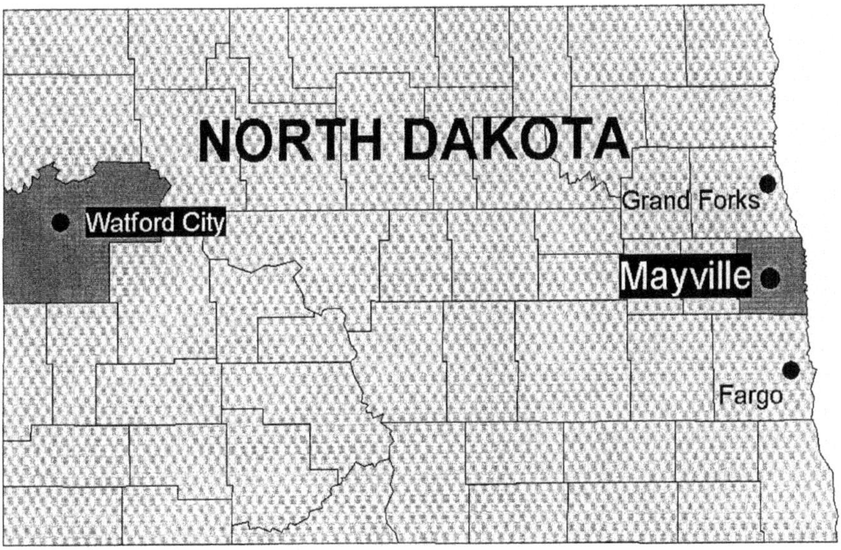

Figure 5-5. Geographic Location of Mayville, North Dakota.

the creation of information technology dependent businesses. In 1998, Mayville State received a Department of Commerce Technology Opportunities Program (TOP) grant to formally establish the incubator and support its daily activities, including staff and facilities management. As a result of the TOP grant, the Trail County Technology Center (TCTC) began in 1999 with a mission to create and retain jobs within the community. According to the TCTC (2002), in January of 2002, the incubator had six tenants, 27 workers, and an estimated annual payroll for participating employees of $625,000.

The most successful of the TCTC businesses is ComMark, a web development company. The company was founded by two college professors and a student in 1995 and currently employs between 12 and 16 employees. It operates out of offices in both Mayville and Grand Forks, North Dakota. ComMark's professional specialties fall into four distinct categories (ComMark 2002):

1) Development and marketing of proprietary products: These include Adjazz, Frostbyte Tickets, and E-Commerce Malls.
2) Contract Programming: ComMark contracts with clients to provide custom programs to specification in Cold Fusion, Active Server Pages, and C++. ComMark has the largest staff of Cold Fusion Certified Technicians in the area.
3) Call Center: ComMark provides and tracks technical support to clients from a call center location in the Mayville offices.
4) Web Hosting and Design: From basic Internet sites to more sophisticated E-Commerce sites.

The success of the ComMark has given Mayville a new hope for growing and diversifying its economic base. Thanks to Mayville being a CERAtown community, a TOP Grant recipient, and home to ComMark, it is an excellent location for case study research.

The interviews reveal valuable information about the role of leadership and policy in the economic development of Mayville. According to survey responses, the representatives are divided on the success of the CERAtown program. Responses are also mixed on the importance of local leadership in the success or failure of the CERAtown initiative. In general, the Mayville case study presents wide-ranging opinions about benefits from the CERA program, the effectiveness of local leadership, and its economic development success. Table 5-4 confirms the diversity of the answers.

In contrast to the accomplishments of the CERA program in Watford City, representatives from Mayville did not universally recognize the local CERAtown initiative as successful. According to the economic development

Table 5-4. Answers to Designated Response Questions—Mayville, North Dakota

Survey Questions	*Economic Development Representative*	*Education Representative*	*Small Business Representative*	*Agricultural/ Service Sector Representative*	*Project Representative*
Are you directly involved in the IT/broadband project?	**YES**	**NO**	**NO**	**NO**	**YES**
Do you view the project successful?	**YES**	**NO**	**YES**	**BOTH**	**YES**
Does the community support the project?	**YES**	**YES**	**YES**	**YES**	**YES**
Did the project increase the economic development activity in the community?	**YES**	**NO**	**YES**	**NO**	**YES**
In your opinion, has the project been driven from the bottom up or top down?	**BOTH**	**NEITHER**	**BOTH**	**BOTTOM-UP**	**TOP-DOWN**
Do you credit local leadership as the most critical factor in the success or failure of the project?	**YES**	**NO**	**NO**	**YES**	**YES**

representative, the CERA project was "somewhat successful." The economic development representative said:

> The most important result of CERA project was the increased technological awareness of the locals. It also helped us in getting the TOP grant. But overall, the plan did not follow through on many of its promises. For example, the Western Governor's Association planned to have someone call on businesses across the country to promote the CERAtowns, but that never happened . . . we were also supposed to receive funding for the project, but that did not happen either. So on one hand, yes it was successful, but on the other hand, not really.

The small business representative thought it was successful, but when asked about the goals of the CERA project the representative said, "I'm not sure what the goals were in the beginning, but I do know that the Internet is now here." The project and agricultural/service representatives said the CERA program was successful, but admitted the project "was not as successful as we would have liked." Yet, the education representative clearly found the CERA project a complete failure. The education representative said:

> We have not seen any activity for the CERA project, not in name or results. The whole project was unclear, with no real leadership from the organization . . . leaving Mayville in a no-win situation . . . It was my understanding that the goal of the CERAtown project was to create jobs, but it didn't. So I would say based on this goal, CERA was a failure.

Based on these responses, the CERA program achieved little success in Mayville.

Just as with inquiries about the success of CERA, responses to questions regarding the importance of local leadership are diverse. The representatives familiar with the CERA project cited top-down leadership as the key factor in the failure of the initiative. The economic development representative said:

> The initial CERA leadership from state government (North Dakota) got the program started. But not long after the beginning of the program, the Governor and Lieutenant Governor were not reelected. The CERA project went down with them. Without this leadership, the CERA responsibilities of the Western Governors' Association disappeared.

The economic development representative backed up this "disappearance" of the CERA leadership through an alarming example. The economic development representative said:

> I can give you an unbelievable example to demonstrate the disappearance of CERA leadership. An older lady in the community called me and said that I needed to take a look at my website. She said that our site had a link to an adult web site and I should remove it. Come to find out it was the CERAtown site. They did not renew the URL address for the site. Now that is what you call abandoning a leadership position.

The economic development representative said that despite the abandonment of the CERA project by the national headquarters, local leadership recognized that a technology-based economic development strategy is the "way to go." The response from the education representative indicates that local leadership plays a vital role in the achievement of quality telecommunications infrastructure. According to the education representative:

> Though the CERA project did not produce jobs, our local leadership made the best out of it. Local leadership did two important things: they made the improvements in infrastructure happen and hired a person to facilitate the technology activities, through the Economic Development Commission's support of the TOP grant application, both in dollars and dedication.

The small business representative did not feel that local leadership was critical for the success or failure of the CERAtown program. The representative felt that all the community needed was the technology; the "enthusiastic" citizens could do the rest on their own and went further to say "small town leaders can hinder more than help."

The cornerstone strategy for local economic development is the technology incubator at Mayville State University. The interviews suggest that local representatives believe the incubator holds the greatest potential for economic development success in Mayville. Despite their confidence in the incubator, additional "problems" were revealed. One apparent problem is the relationship between Mayville and the neighboring community of Portland. Unlike Watford City, the only substantial town in McKenzie County, Traill County is home to several communities of comparable size. The need to "spread the wealth" of economic development around the county seems to take away the growth potential of Mayville.

In general, the Mayville case study did not prove itself a model of economic development "success." Despite Mayville's strong institutions and favorable geography, its limited "success" underscores the importance of both local leadership and economic development policy. But the lack of success within the CERA program does not discount the significance of Mayville's research contribution. In fact, the limited economic development activity of

Mayville emphasizes the role of "intangibles" for the economic success of rural communities.

In closing, the case studies reveal three fundamental concepts important for broadband telecommunications-based economic development success: local leadership, technical assistance, and quality-of-life economic development planning initiatives.

1) Local leadership is a vital component to rural telecommunications-based economic development. Though leadership is important for all types of economic development activities, it is even more important in technology-based initiatives because of the need to convince rural citizens of its significance for economic growth and development. As exemplified by Mayville, the absence of leadership can limit the possibility of local economic development success. The research also reveals that leadership is most effective when it comes from both the top-down and bottom-up.
2) Local technical assistance is required for moving the community toward utilization of information and Internet technologies. In addition, the cases reveal how "local" help, like high school students, encourages participation and reduces the risk of trying the technology. The combination of technical assistance and leadership strategies contribute to the success of the Ennis "Information Age" project.
3) Economic development strategies must focus on quality-of-life improvements, not attracting industries. Watford City exemplifies how quality of life economic development planning and advanced telecommunications can work for job creation.

The following section discusses these concepts further and incorporates each into the structural model for broadband-based rural economic development planning.

UTILIZING BROADBAND TELECOMMUNICATIONS FOR ECONOMIC DEVELOPMENT: A STRUCTURAL MODEL

The overall purpose of this research is to develop a practice-oriented model to assist the economic development community in applying broadband telecommunications for rural job creation. Though the chapters contain academically accepted research methods, this structural model is directed toward "real-world" applications for local and state economic development officials. The term "structural model" refers the general framework for initiating rural economic development planning based on broadband telecommunications. The structural model, though quite basic, presents the necessary components for broadband telecommunications-based rural economic development planning. The rationale of the model is to provide a framework for guiding economic development planning while using the preceding research chapters for descriptive reference. In essence, the reader could begin with the model and return to previous research chapters for further details of ideas and techniques to insert into the practice-based model. For example, when the model calls for the development of broadband access policy, the user can refer to Chapter Three for a detailed presentation of state policies.

As noted earlier in the research, the basis of this framework is clusters. Accordingly, the research findings from the previous chapters serve as the basis of the model. In order to properly "set the stage" for the development of a model, the research findings are revisited.

Chapter Two establishes the importance of broadband telecommunications as the "Interstate Highway of the 21st Century." Within this chapter, broadband telecommunications is recognized as the transportation network of the new economy. Without access to broadband, rural economies are limited in their ability to participate and compete within the context of the information economy.

Chapter Three examines federal and state policies related to both rural broadband access and economic development. This chapter lays the foundation for state-level broadband-based rural economic development strategies. The development of a rural-based model is dependent on the environment in which communities must compete. Based on the content of the federal and state policies, innovative approaches to rural access and economic development initiatives are discovered. The model incorporates these innovative policy findings of Chapter Three.

Chapter Four uses factor analysis to recognize rural locations with the potential to develop rural technology-based clusters. Using the CERA communities as the focus of study, a "broadband development index" ranks the prospect of the study area counties for rural clusters. Though the details of the statistical techniques may not be of interest to economic developers, the resulting factor scores, index, and maps contribute significantly to the development of state, regional, and local economic development planning. With sufficient data, the factor analysis approach to identifying rural clusters transfers easily to any study area.

Chapter Five observes the importance of local leadership and economic development planning for information technology-related job creation in rural communities. The qualitative research provides information about the economic development "intangibles" not easily identified on the surface. This chapter introduces local phenomenon, to build on quantifiable factors identified in Chapter Four. The case studies reveal three fundamental concepts important for broadband telecommunications-based economic development success: local leadership, technical assistance, and quality-of-life economic development planning initiatives.

Before incorporating the results of the research chapters, it is necessary to establish the structural framework of the model. The structure of the model depends on two significant theoretical approaches: 1) rural knowledge clusters and 2) leadership compression.

Rural knowledge clusters focus on the following dynamics important to technology-based economic development: leadership, location, labor, economic and social characteristics, higher education and research institutions, and technological capacity. As noted in Chapter Four, the application of rural knowledge clusters is appropriate for this model because the capacity to create knowledge-based businesses and employ educated workers is essential for economic development success within the new economy.

The majority of American rural economic development strategies are "top down" in nature (Okagaki, et. al 1999). The preceding chapters support this belief. For example, Chapter Three confirms the top-down structure of rural broadband telecommunications infrastructure policies. Though top-down strategies are prevalent in rural economic development planning, the literature recognizes bottom-up or local strategies and initiatives as critical for successful economic development research and practice (Callanan 2000; Edwards 1996; Grant 1990).

The advancement of rural economic development research depends on methodologies incorporating the experiences of real communities. James Grant said:

> The intervention of governments to help under-developed or distressed areas is generally based on ideology, or political pressure, or on someone's hunch. We need a much more scientific approach to these matters, based on practical experience in the field (Grant 1990, 21).

Using practical experience as a guide is the only way to advance both the research and practice of economic development. According to Edwards:

> It is impossible to understand real-life problems unless we grasp the multitude of constraints, imperfections, and emotions which shape the actions of real people. Conventional research cannot do this because it divorces itself from the everyday context in which an understanding of these emotions can develop. We (development researchers) cannot be relevant to people unless we understand their problems, but we cannot understand these problems unless people tell us about them (Edwards 1996, 78–79).

Also, the case study research in Ennis, Ireland supports the significance of top-down and bottom-up strategies. Based on these findings, the model framework utilizes a leadership compression approach, which is a combination of top-down and bottom-up techniques for leadership and economic development planning. The two paradigms of leadership compression and rural knowledge clusters offer useful approaches for the construction of a rural economic development-planning model based on broadband telecommunications.

The model construction takes place in three distinct stages: 1) division of state and local activities, 2) step-by-step introduction of strategies, and 3) leadership structure.

The first stage in the model building is the separation of state and local economic development activities. This is not to suggest that state and local economic development activities are mutually exclusive, but rather to create a framework where the introduction of model components is more understandable. Figure 5-6 presents the first stage of model construction.

The second stage of model construction is a step-by-step introduction of the policies and schemes for rural economic development based on broadband telecommunications. This second stage constitutes the bulk of the model. The research results from Chapters Two through Five serve as the basis of strategy development. The step-by-step strategies are divided by state and local entities.

The strategic components of the model begin with the introduction of state strategies. The strategic steps are the following:

1) Step One: Inventory of Broadband Telecommunications
2) Step Two: Policy Development: Access-focused Policy
3) Step Three: Identification of Potential Rural Clusters
4) Step Four: Policy Development: Economic Development Planning
5) Step Five: Connecting with Local Communities

The first step in state-level strategy is to inventory the broadband telecommunications infrastructure. This includes telecommunications, cable, and alternative providers, such as electric cooperatives. A plan of action cannot take place without this inventory. The inventory is vital for economic development planning. As exemplified in Chapter Three, states such as North Carolina, Idaho, and Pennsylvania made use of the inventory to better understand the current level of service to rural communities. In short, without knowing what you have, you cannot determine what you need.

Once the inventory is available, assessing rural access to broadband is the second step. The research suggests a group of policy approaches to deal with

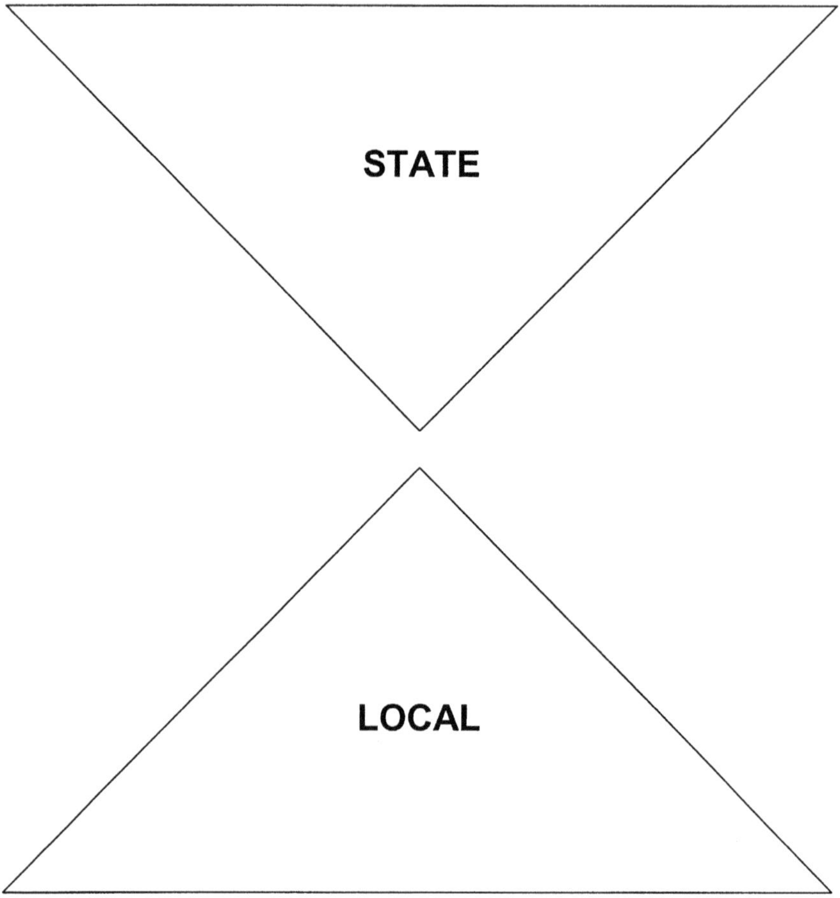

Figure 5-6. Step One of the Structural Model: Division of State and Local Economic Development Activities.

the lack of broadband telecommunications and other information technologies in rural areas. The major policy concern is closing the rural access gap. This involves the identification of private suppliers and the development of innovative strategies, e.g. broadband tax credits, grant and loan programs, etc., to encourage broadband deployment and consumer utilization. These approaches also include the identification of federal programs that assist rural areas in gaining broadband access.

Step three is the development of economic development specific policies for technology-based economic development. These legislative proposals are innovative approaches ranging from rural strategic planning to technology outreach programs. These policies offer great potential to assist rural communities in acquiring and utilizing broadband telecommunications for job creation. The best method to develop innovative policies is to borrow from other states that have created rural economic development through broadband accessibility. According to the research, these progressive states include North Carolina, Georgia, and Idaho. The research in Chapter Three supports the importance of steps one, two, and three for state level economic development planning. For additional information in the development of these steps, please reference Chapter Three.

The identification of potential rural clusters is step four. Based on the literature, clusters are one of the most popular state-level economic development strategies. This research offers a methodology to identify potential clusters in rural communities. It is important to note that not all rural communities have the ability to develop clusters. State-level economic development representatives must give attention to towns with the capacity to build clusters. Based on the factor analysis, these rural towns tend to be sub-regional trade centers, usually offering services that attract neighboring communities. Such attractions include shopping, universities and colleges, and service employment. Though numerous cluster identification methods exist, Chapter Four provides a useful tool for identifying and ranking counties with the potential to develop rural clusters.

Step five is the final state-level piece of the model. This is the connection with local communities to determine ways to coordinate state and local economic development efforts. The contact also assists the state official in determining the awareness and capabilities of the local leadership. It is possible that a rural community does not know that cluster-based development strategy would be appropriate to create local jobs. This is the top-down leadership role noted within the case studies of Chapter Five. With the exception of LaGrange, top-down assistance proved valuable in the establishment of broadband-based economic development programs. As shown in Figure 5-7, the model incorporates these steps into the area designated for state economic development activities.

It is now time to introduce the local initiatives in the construction of the model. The steps are:

1) Step One: Connect with State Economic Development Leadership
2) Step Two: Conduct local broadband access and demand study
3) Step Three: Develop a "quality of life" development plan
4) Step Four: Develop a "local technical assistance" program
5) Step Five: Initiate interest in information technologies through local programs

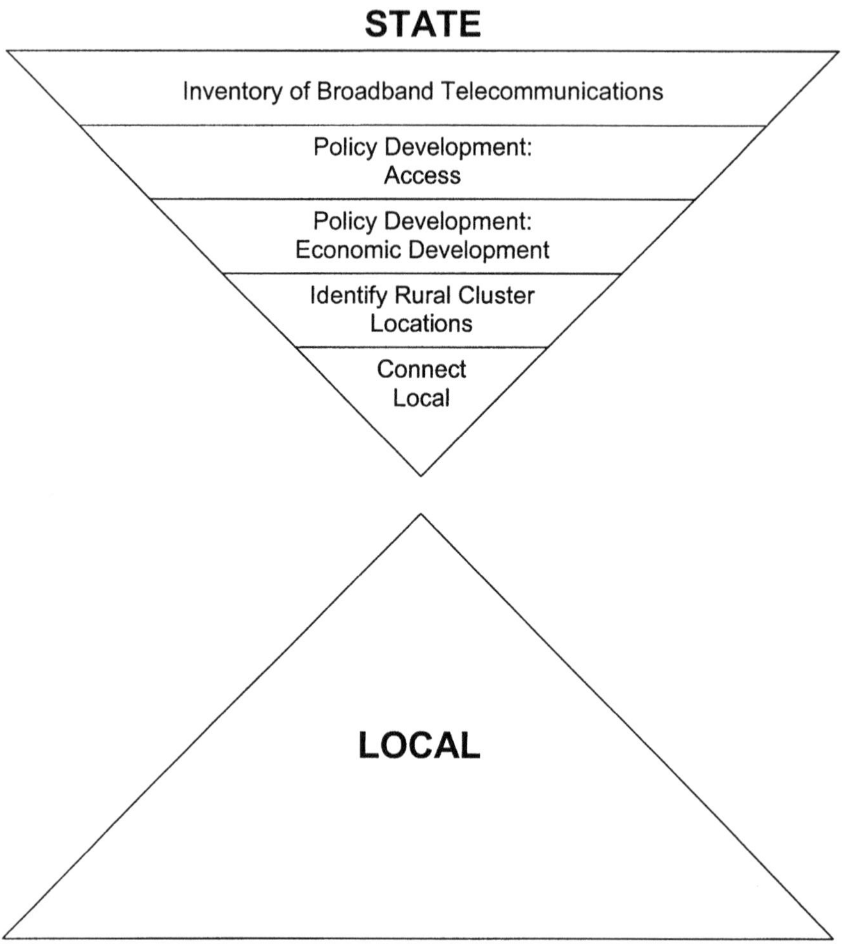

Figure 5-7. Step Two of the Structural Model: Introduction of State Economic Development Strategies.

The first step in local strategy is contact with state-level economic development officials. For the local economic development official this is an excellent way to determine if state leadership is in touch with the needs of rural communities. More importantly, state offices can provide information about state-level economic development goals and strategies. It is important to note that contacting the local representative is the last step for the state, but the first step for local. As exemplified with Chapters Three and Five, leadership and policy directives generally move downward. Thus, it is logical that state leaders should be more prepared to address the concerns of local economic development officials.

In step two, it is the job of local economic development representatives to inventory the broadband access and demand needs of the area. Access inventory is a simple study process, you either have some form of broadband or you do not. The demand study gives the local leadership a better understanding about broadband demand from businesses, educational institutions, individuals, etc. The demand response speaks volumes about the potential success of a broadband development strategy for the community. This is the point where the economic developer can have a significant top-down leadership influence. The economic developer can inform the community about possibilities and opportunities through broadband connectivity. Chapter Two points out the importance of connectivity within the context of the new economy through numerous examples.

Step three is the development of a "quality-of-life" focused strategy for rural economic development. Instead of chasing industry, the focus of local economic development officials is making the community attractive to existing citizens with the hopes of attracting newcomers. Watford City uses this approach to make life better for locals, while at the same time enticing new businesses and people to town. It only makes sense that if broadband technology enables a person to work anywhere, naturally people will live and work where quality of life is high. The focus of economic development officials in Watford City moved away from the industrial park to movie theaters and restaurants.

The next step in the model is the development of a local technical assistance program. This policy initiative gives rural citizens the confidence to try computer and Internet technologies without fear of failure. Ennis, Ireland utilizes local high school students to provide local technical assistance for first time users. This strategy requires little expense, yet proves valuable to increase computer ownership and Internet subscription demand. Earlier results from this chapter serve as the research basis for inclusion in the model.

The final step in local activities is the establishment of local technology utilization programs. In short, these programs encourage the use of information technologies by locals. This strategy introduces a bottom-up approach, which promotes ownership and participation in technology-based economic development strategies. Once again, case study research from Chapter Five reveals numerous strategies. For example, these strategies offer local church bulletins and Sunday school lessons over the Internet (Ennis), make Microsoft training and certification available (Watford City), and throw occupation-specific parties to encourage citizens to buy a computer (Ennis). Figure 5-8 presents the incorporation of local economic development strategies into the model.

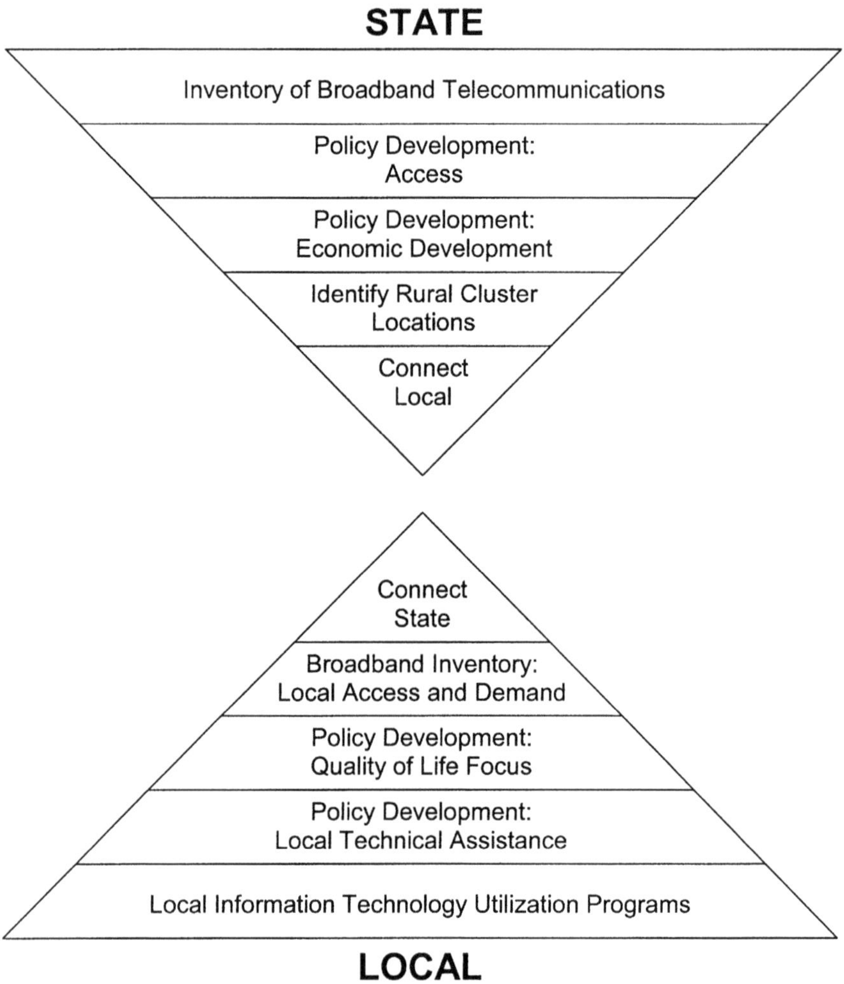

Figure 5-8. Step Two of the Structural Model: Introduction of Local Economic Development Strategies.

The final piece of the model recognizes the importance of leadership compression for economic development success. By encouraging the simultaneous top-down and bottom-up leadership initiatives, rural towns, such as Ennis, Ireland and Watford City, North Dakota, achieve economic development success using broadband telecommunication and information technologies as the foundation. These leadership compression forces are presented in the model with colored arrows, red for bottom-up and black for top-down leadership forces. Since this is the final step in the construction of the model, Figure

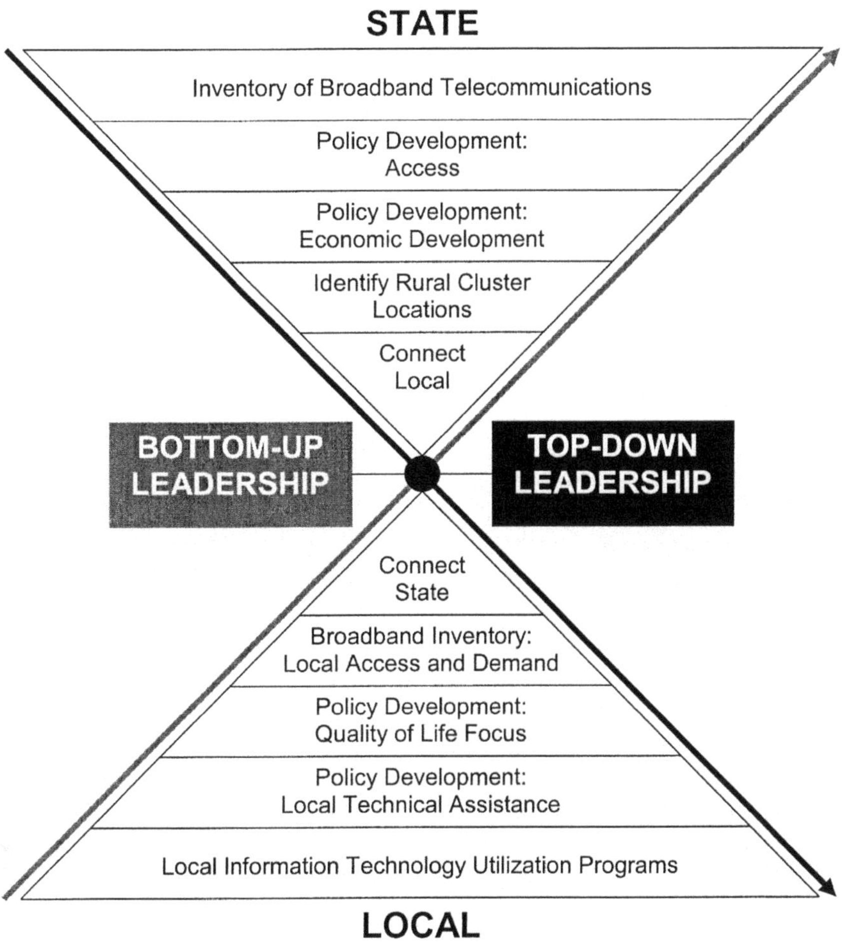

Figure 5-9. The Structural Model for Broadband Telecommunications-Based Rural Economic Development Planning.

5-9 represents the complete structural model for broadband-based rural economic development planning.

In conclusion, the structural model provides a foundation for the rural economic development professional interested in broadband-based rural economic development. It builds on the research in such a way that pieces of study can be utilized directly in rural economic development practice. Though intended for local practitioners, meso and macro level economic developers will benefit from both the survey of broadband policies and the rural cluster identification technique.

Chapter Six

Conclusion

As suggested by the title of this research, many rural communities need "deliverance" from economic despair. The definition offered in Chapter One suggests that deliverance is "the act of giving up or handing over; liberate and rescue" (Oxford English Dictionary 1993, 626). Another key piece of the title is the idea of "dragging rural America kicking and screaming into the information economy." This portion of the title reflects both the perceived notion that rural communities are reluctant to accept change and the disconnectedness of these areas to emerging economic and technological trends. Being from a small town, the problem of economic survival without adaptation to emerging trends and technologies is very real. In order to survive in the new economy, rural communities must begin to accept continuous change as normal. Rural America must embrace the possibilities and pitfalls of the new economy if they wish to stay alive. Whether rural areas get there alone or dragged "kicking and screaming," for many remote areas the economic possibilities offered by information technologies and advanced telecommunications cannot be ignored.

Accordingly, this research contributes to the understanding of how advanced telecommunications and information technologies can be utilized to create jobs in rural communities. More importantly, the findings of this study are used to construct a structural model providing a generalized framework to initiate rural economic development planning through the application of broadband telecommunications.

LESSONS FROM THE RESEARCH

This chapter offers conclusions on the possibility of creating rural economic development opportunities through the application of broadband telecommunications and proposes future research directions. Based on the preceding research chapters, the results generate four significant conclusions:

1) Broadband telecommunications can improve the economic development opportunities of rural areas.
2) Access to broadband communications alone does not result in rural economic development success.
3) Although broadband access is vital, economic development success depends on factors that encourage broadband and information technology utilization, including leadership, state and local economic development policy, and technical assistance.
4) Clusters are an acceptable strategy for rural economic development in communities that have sub-regional trade center traits, such as institutions of higher education, shopping facilities, service employment opportunities, etc.

The first significant conclusion is that broadband telecommunications can improve the economic development opportunities of rural areas. Broadband telecommunications are a critical component for linking business activities in the "new" economy. Thus, just as the interstate highway played an important role for movement of goods within the "old" economy, broadband telecommunications serve as the transportation system of the "new," information-based economy. This finding is significant for rural communities. As noted in Chapters One and Two, economic development activity in rural communities traditionally suffers because of their remote geographic location. This technology eliminates much of that remoteness, but at the same time requires quality pools of labor and high quality-of-life amenities to compete with other places.

Though the importance of geographic location decreases, new challenges emerge for rural areas because of low population and limited quality-of-life opportunities. This situation makes competing with urban areas even more difficult. Despite the challenges, it does open the door for some rural communities who have the ability to develop clusters through strong institutions and capable leadership. The case studies presented in this research prove that broadband telecommunications can position rural communities for growth, but moreover, awaken rural inhabitants to the reality that technology is here to stay. As with Watford City, North Dakota, the citizens moved beyond the mere acceptance of technological change, they embraced its promise.

Fortunately for rural Americans, federal and state policymakers recognize the need for assuring broadband access to remote areas. This research identifies several trends in federal and state policy with regard to broadband access for rural regions. From a federal perspective, the private sector is doing an acceptable job of supplying broadband telecommunications to rural areas.

Though to insure its timely roll out, proactive federal policies include tax cuts for rural telecommunications investment and low interest loan programs to speed deployment. Despite the proactive policy, the most apparent function of federal policy is assuring that competition remains fair among telecommunications companies and cable television providers. In addition to federal concerns, the research suggests that state legislative and executive policies provide additional support for supplying rural communities with broadband telecommunications. From the analysis of state policy, two significant conclusions are recognized with regard to strategies for encouraging broadband access. First, assuring rural access to broadband telecommunications begins with an inventory of exiting infrastructure. Secondly, states with concerns about timely access offer alternative approaches to supplying broadband. For instance, states such as Nebraska and North Carolina have aggressively passed legislation and initiated programs to further expand digital communications technologies for rural citizens, even going as far as providing access to rural communities via the state backbone.

The second major conclusion is that access to broadband telecommunications alone does not create rural jobs. Broadband telecommunications can be like a speculative building in an industrial park, available but not utilized. Thus, telecommunication-based development, just as with any other traditional economic development strategy, operates in an environment where not all locations succeed in creating local job opportunities. Endogenous traits, such as interstate access, universities, access to rivers, skilled labor, etc., make the economic development situation for some communities more profitable than others.

Although broadband telecommunications are a crucial component for the efficient movement of information in the global economy, not every location is equal in its ability to use it. Yet, rural communities without broadband telecommunications will face serious problems in the future. In particular, the fate of many existing rural businesses depends on connectivity. As noted in Chapter Two, the emergence of e-commerce is changing the structural basis for business transactions. For rural businesses that export goods and services, acknowledging this trend is imperative. The findings of Chapters Four and Five also support the idea that planned efforts, such as the CERA program and the eircom "Information Age Town," are needed to spur local utilization of technology. This is not to say that utilization efforts must be on such a large scale, but rather to simply recognize that deliberate efforts, no matter the size, are needed to encourage technology use by rural residents.

The third significant conclusion is that economic development success depends on factors that encourage broadband and information technology utilization. This conclusion supports the preceding idea of planned utilization programs, but more specifically identifies issues that contribute to rural economic development success. The case study research in Chapter Five reveals three fundamental concepts important for broadband telecommunications-based economic development success: local leadership, technical assistance, and quality-of-life economic development planning initiatives.

Local leadership is a vital component to rural telecommunications-based economic development. Though leadership is important for all types of economic development activities, it is even more important in technology-based initiatives because of the need to convince rural citizens of its significance for economic growth and development. As exemplified by Mayville, the absence of leadership can limit the possibility of local economic development success. The research also reveals that leadership is most effective when it comes from both the top-down and bottom-up, recognized in the research as "leadership compression."

In addition to local leadership, local technical assistance is necessary for moving the community toward utilization of information and Internet technologies. The case studies reveal how technical assistance provided by local high school students encouraged participation and reduced the risk of testing the technology. The importance of both technical assistance and leadership is exemplified through the success of Ennis, Ireland and Watford City, North Dakota.

Perhaps the most noteworthy conclusion of this research is the need for rural communities to refocus the local economic development planning efforts from recruiting to quality-of-life improvement. Watford City exemplifies how quality-of-life economic development planning and advanced telecommunications can work for job creation. As noted in the interviews, Watford City economic development officials recognized that its geographic location made bringing in outside industry dependent on transportation impossible. Thus, local economic development efforts turned from recruiting to focusing on local quality-of-life issues. This research brings to light the need for rural economic developers to move away from recruitment strategies and focus on improving the livability of the local community.

The fourth and final conclusion from this research is the recognition of clusters as a valid strategy for rural economic development. Based on both the review of literature and results from the factor analysis in Chapter Four, it is evident that certain economic, demographic, and geographic factors are important in the implementation of information technologies and broadband telecommunications. Yet, it is also clear that cluster development is not an ideal strategy for every rural community. The development of clusters depends on the initial ability of a rural community to retain and attract "talented" labor through inherent institutional advantages, such as institutions of higher educations, retail sales outlets, and service employment opportunities.

DIRECTIONS FOR FURTHER RESEARCH

Though this research does advance the understanding of telecommunications-based rural economic development, it is far from complete. At numerous points during the study, further research considerations appeared. Of these many ideas, three significant research topics need further investigation.

First, research is needed in rural communities that have access to broadband telecommunications, but are without local economic development assistance. In general, it is accepted that communities have a local economic development official, but that is not the case for many rural areas. In contrast to my research in mostly successful communities, studies in localities without any local development assistance would add or subtract the importance of economic developers in the process of creating jobs through broadband technology.

Second, the "broadband development index" presented in this research needs further refinement. This tool holds great potential in the identification of rural communities with cluster potential. Yet, further work is needed to identify variables contributing to the detection of rural communities with development possibilities through cluster-building strategies. In particular, developing quantitative approaches to quality-of-life indicators, which proved most influential in Watford City. If this index approach could be further refined, it would prove valuable to state and federal economic development leaders seeking to develop strategic plans for rural economic development.

Finally, more study is required to better understand the influences of local, state, and federal policies in the deployment and utilization of broadband telecommunications. This research approach should focus on the private suppliers of broadband telecommunications and businesses that require broadband capabilities to transfer information services. In short, examine the private side of the broadband equation. If researchers can better understand the needs of these businesses, policy recommendations can be formulated to assist in all levels of economic development-related planning.

In conclusion, this research achieves its initial purpose, to develop a structural framework for guiding broadband-based rural economic development planning. Along the way, the difficulty of achieving rural economic development success becomes quite clear. In general, the practice of economic development is comparable to running a marathon because it is long and requires plenty of stamina. If economic development can be characterized this way, then rural economic development is best described as a marathon with hurdles. To the point, economic development success is hard enough to achieve, but rural economic development is even more grueling. It is my hope that this research assists in removing at least one of the many hurdles facing rural communities in the pursuit of economic development success.

Appendix A

Factor Analysis Variables

GEOGRAPHIC AND RAW DATA PRESENTATIONS

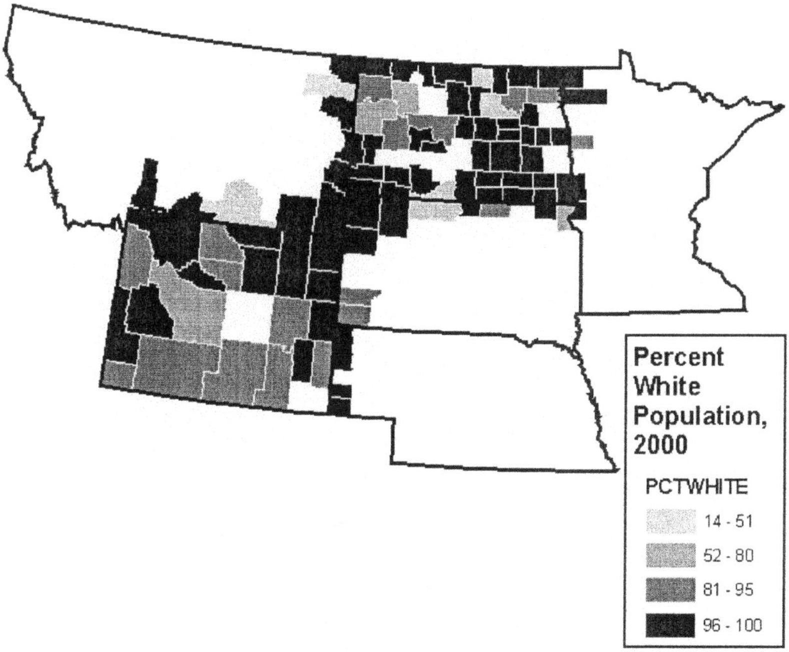
Percent
White
Population,
2000
PCTWHITE
14 - 51
52 - 80
81 - 95
96 - 100

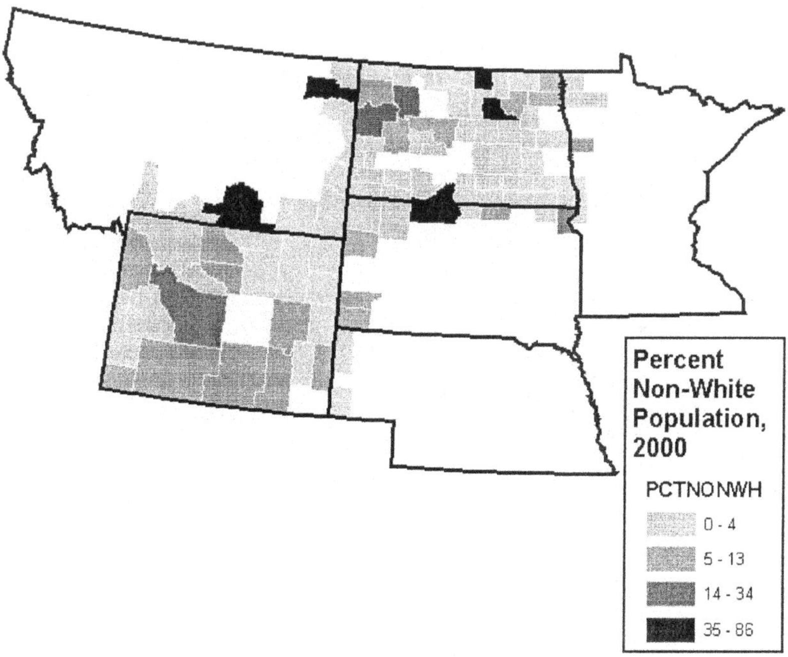
Percent
Non-White
Population,
2000
PCTNONWH
0 - 4
5 - 13
14 - 34
35 - 86

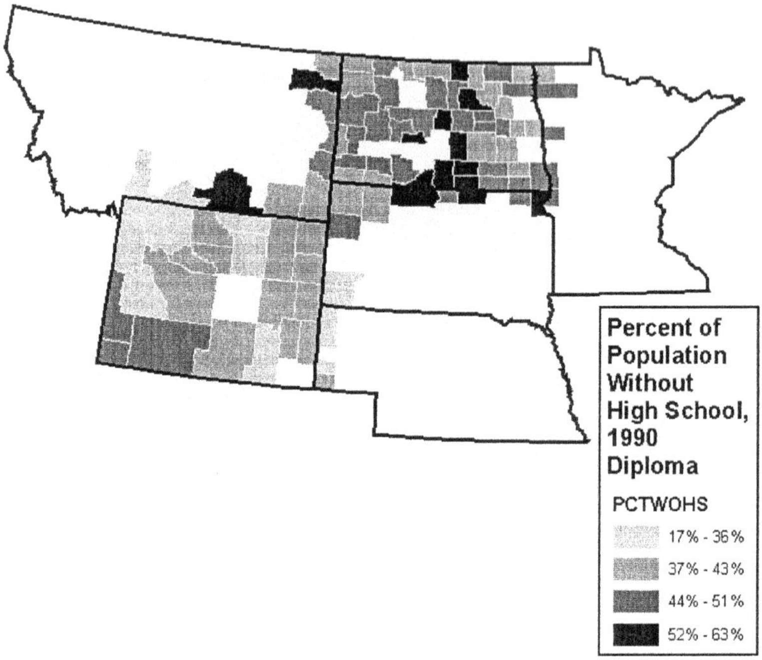
Percent of Population Without High School, 1990 Diploma
PCTWOHS
17% - 36%
37% - 43%
44% - 51%
52% - 63%

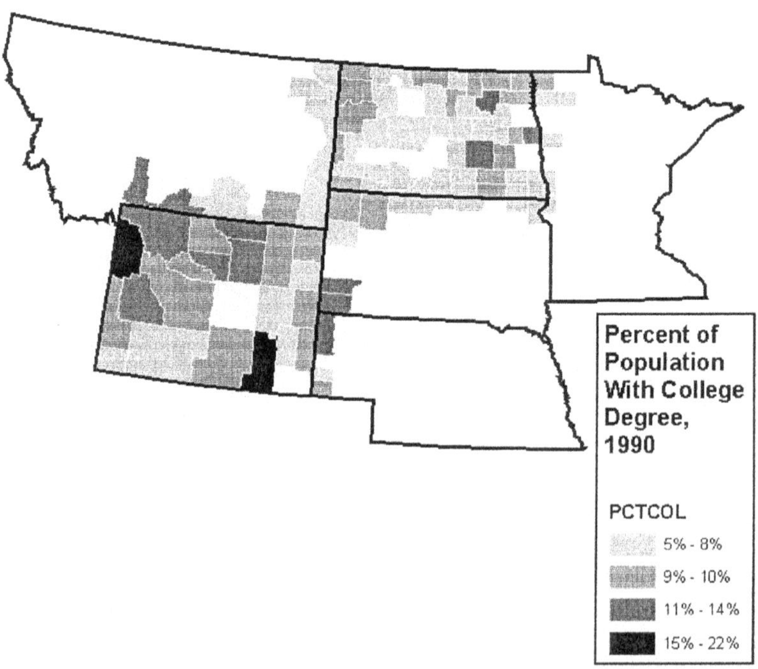
Percent of Population With College Degree, 1990
PCTCOL
5% - 8%
9% - 10%
11% - 14%
15% - 22%

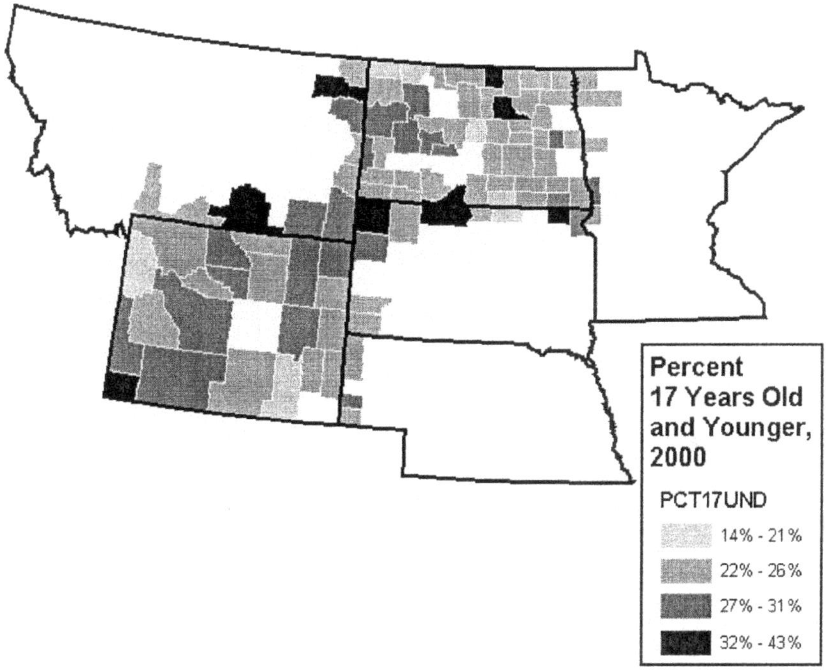
Percent
17 Years Old
and Younger,
2000
PCT17UND
14% - 21%
22% - 26%
27% - 31%
32% - 43%

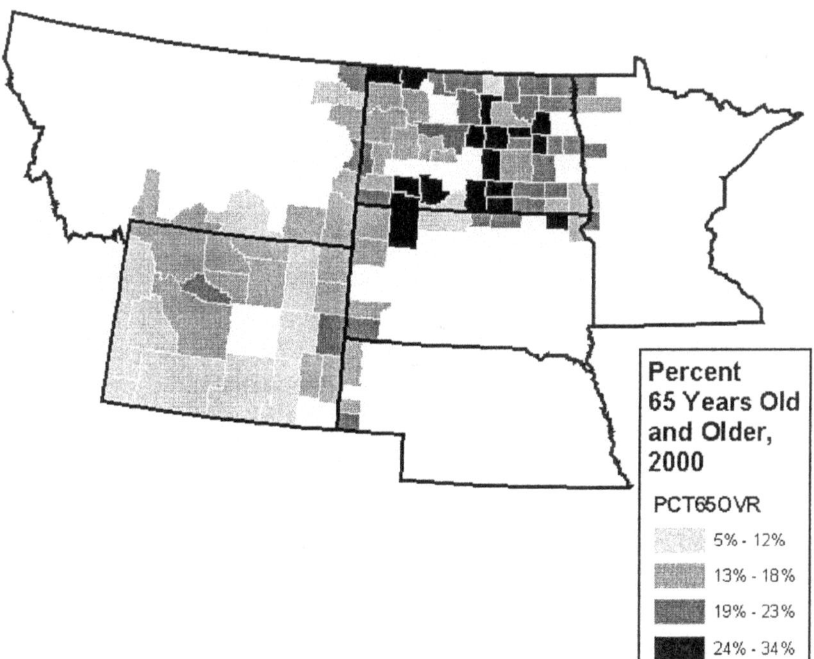
Percent
65 Years Old
and Older,
2000
PCT65OVR
5% - 12%
13% - 18%
19% - 23%
24% - 34%

Median Age, 2000
MAGE2000
24 - 34
35 - 40
41 - 44
45 - 51

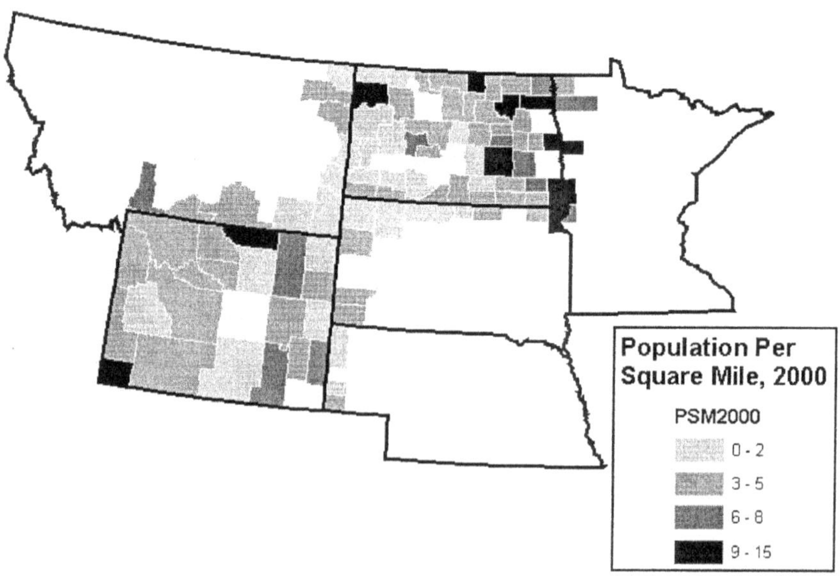
Population Per Square Mile, 2000
PSM2000
0 - 2
3 - 5
6 - 8
9 - 15

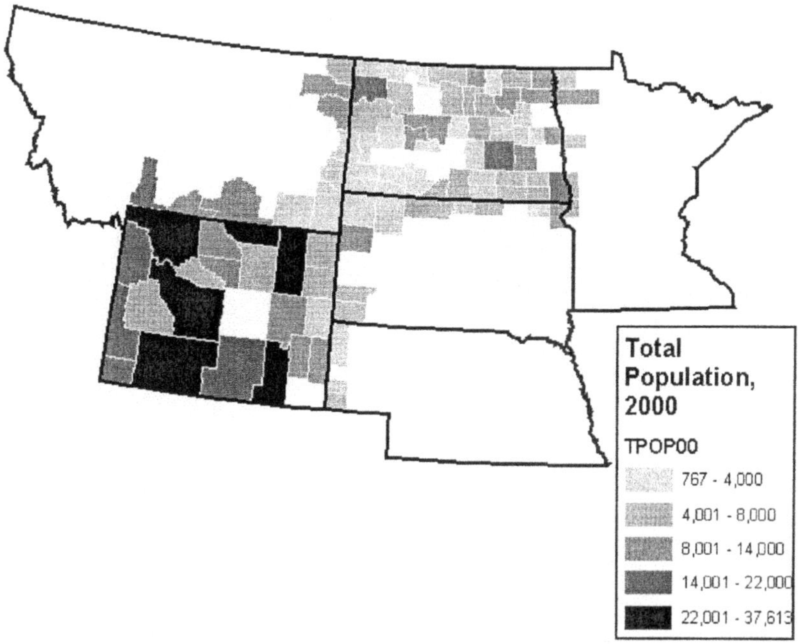
Total Population, 2000
TPOP00
767 - 4,000
4,001 - 8,000
8,001 - 14,000
14,001 - 22,000
22,001 - 37,613

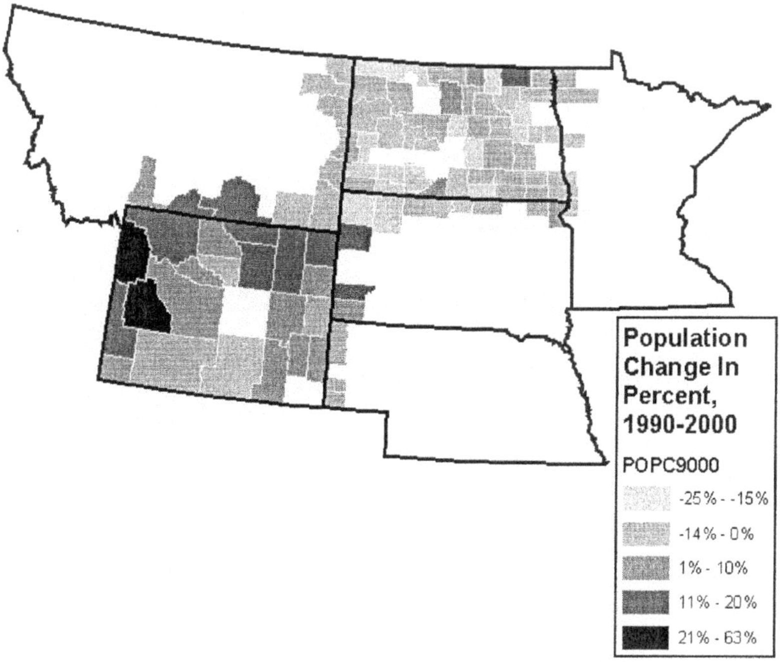
Population Change In Percent, 1990-2000
POPC9000
-25% - -15%
-14% - 0%
1% - 10%
11% - 20%
21% - 63%

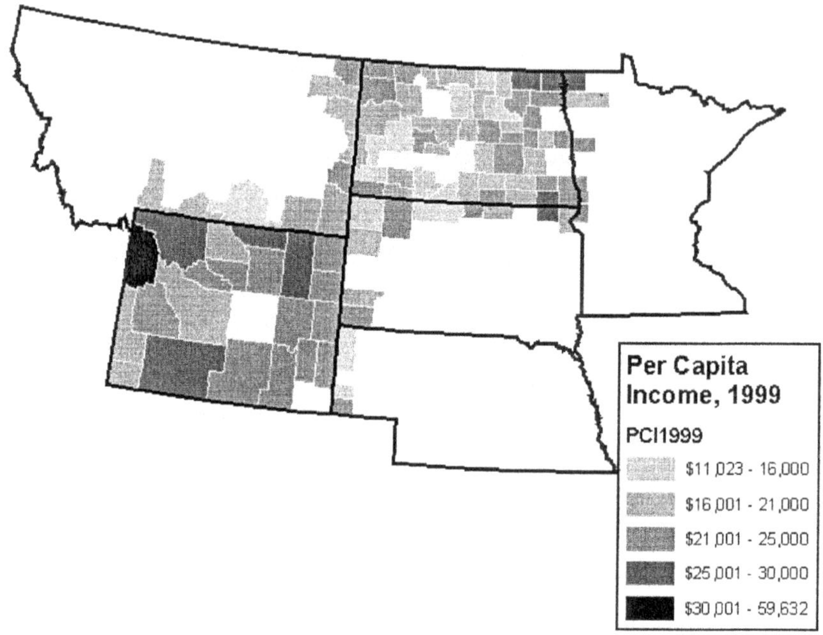
Per Capita Income, 1999
PCI1999
$11,023 - 16,000
$16,001 - 21,000
$21,001 - 25,000
$25,001 - 30,000
$30,001 - 59,632

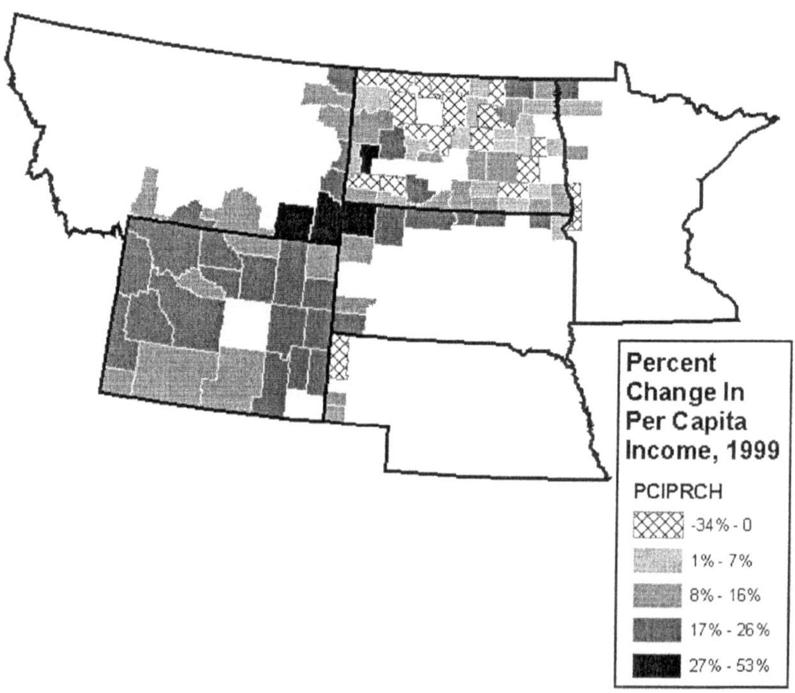
Percent Change In Per Capita Income, 1999
PCIPRCH
-34% - 0
1% - 7%
8% - 16%
17% - 26%
27% - 53%

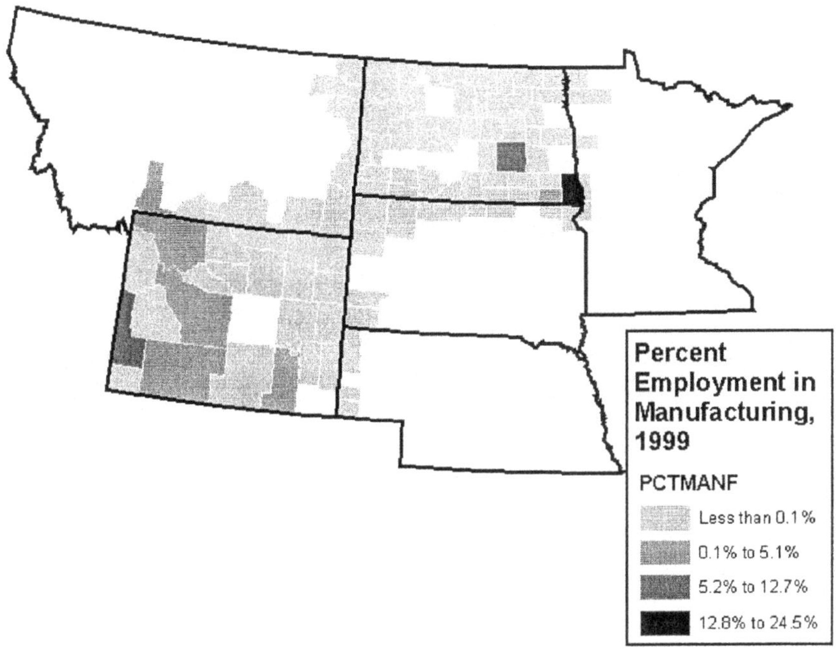
Percent Employment in Manufacturing, 1999
PCTMANF
Less than 0.1%
0.1% to 5.1%
5.2% to 12.7%
12.8% to 24.5%

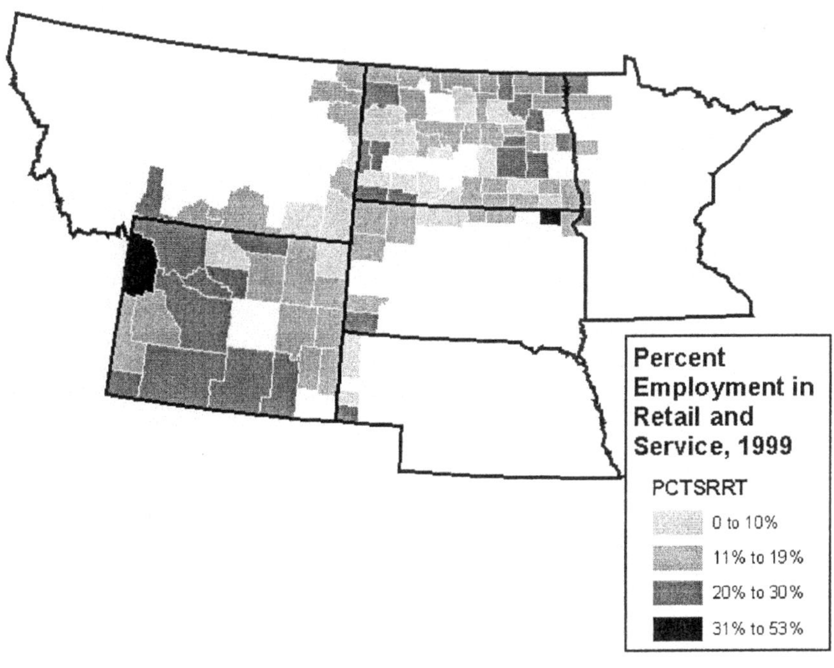
Percent Employment in Retail and Service, 1999
PCTSRRT
0 to 10%
11% to 19%
20% to 30%
31% to 53%

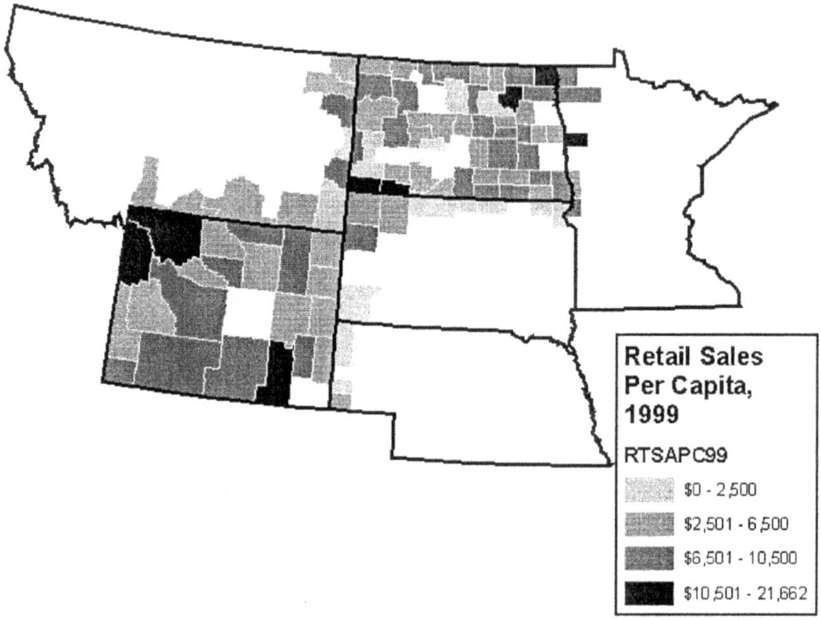
Retail Sales Per Capita, 1999
RTSAPC99
$0 - 2,500
$2,501 - 6,500
$6,501 - 10,500
$10,501 - 21,662

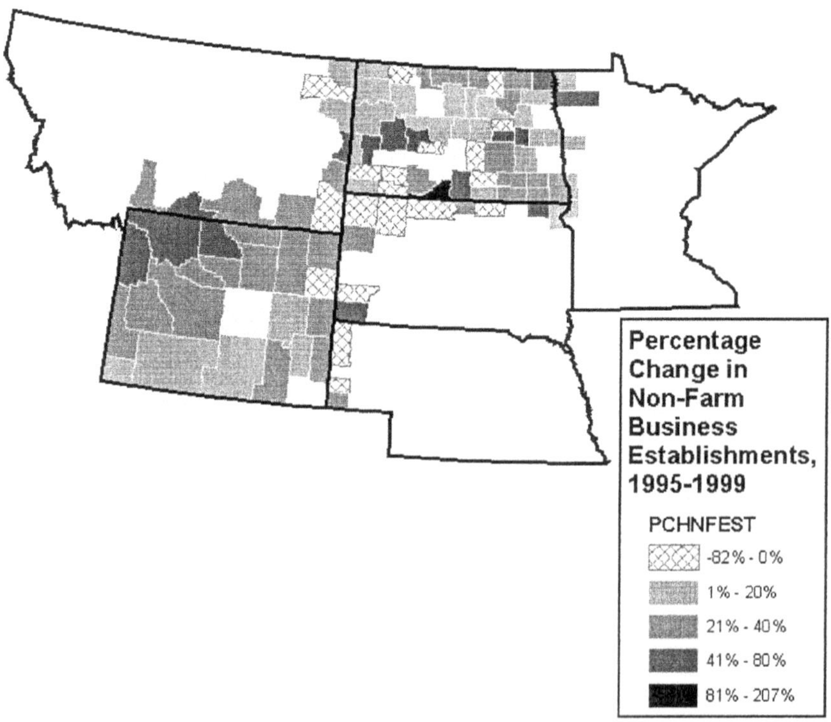
Percentage Change in Non-Farm Business Establishments, 1995-1999
PCHNFEST
-82% - 0%
1% - 20%
21% - 40%
41% - 80%
81% - 207%

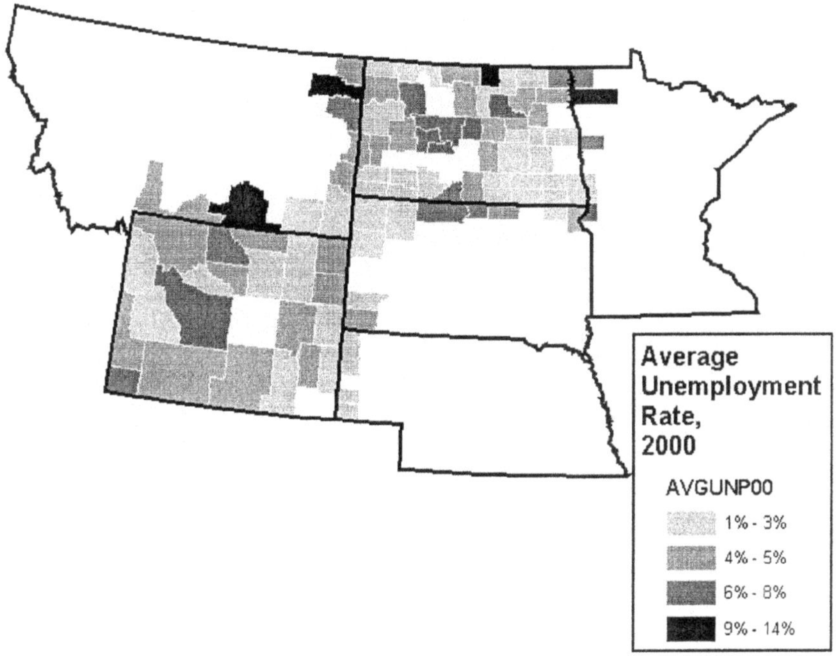
Average Unemployment Rate, 2000
AVGUNP00
1% - 3%
4% - 5%
6% - 8%
9% - 14%

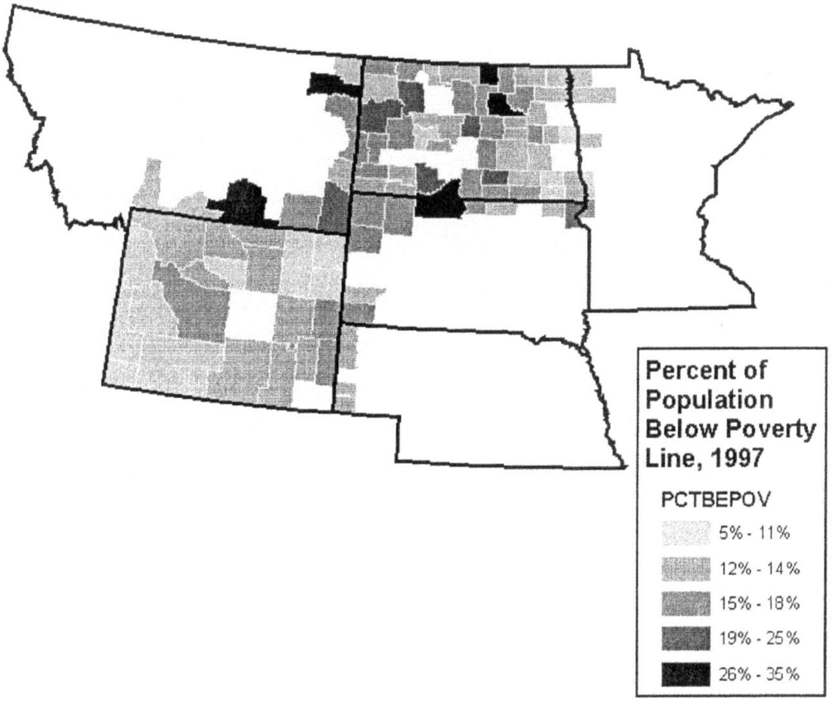
Percent of Population Below Poverty Line, 1997
PCTBEPOV
5% - 11%
12% - 14%
15% - 18%
19% - 25%
26% - 35%

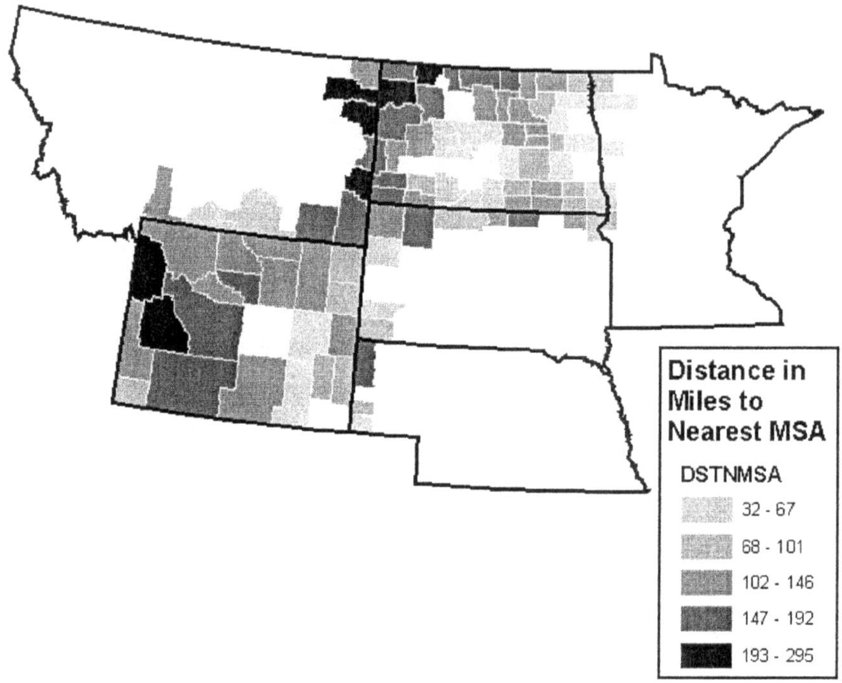
Distance in Miles to Nearest MSA
DSTNMSA
32 - 67
68 - 101
102 - 146
147 - 192
193 - 295

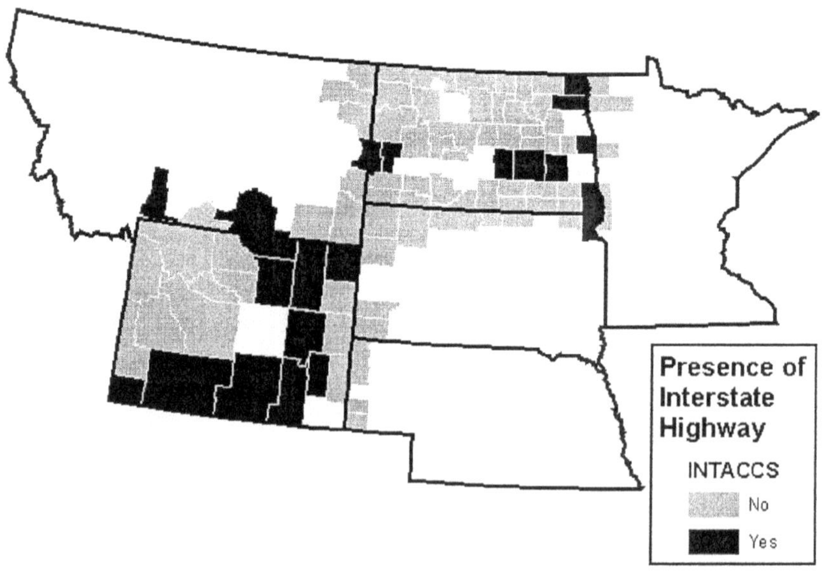
Presence of Interstate Highway
INTACCS
No
Yes

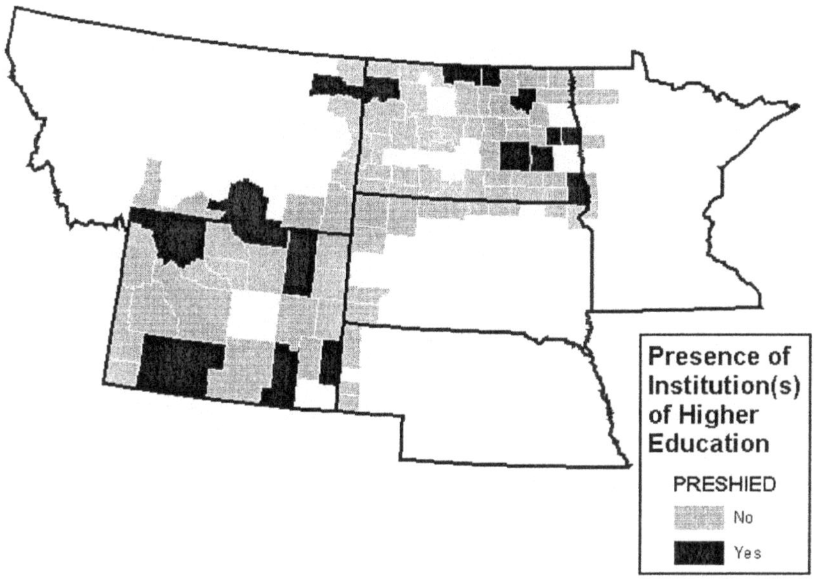
Presence of Institution(s) of Higher Education
PRESHIED
No
Yes

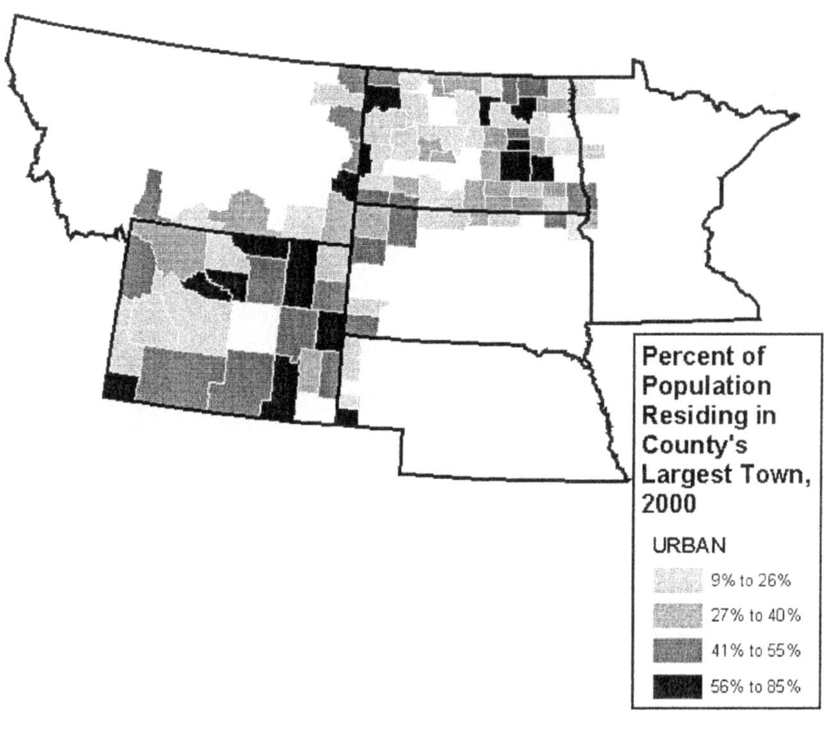
Percent of Population Residing in County's Largest Town, 2000
URBAN
9% to 26%
27% to 40%
41% to 55%
56% to 85%

RAW DATA TABLES

County/State	mage2000	psm2000	popc9000	tpop00	pctwhite	pctnonwh	pct17und	pct65ovr	pci1999	pciprch	rtsapc99	avgunp00
Adams ND	45.6	2.6	-18.3	2593	98.5	1.5	23.2	24.1	18982	2.3	12366	2.5
Albany WY	26.7	7.4	4	32014	91.3	8.7	18.4	8.3	24239	23.1	11415	1.6
Banner NE	39.9	1.1	-3.9	819	95.8	4.2	28.8	15.3	15892	8.1	0	2.8
Barnes ND	40.6	7.8	-6.1	11775	97.9	2.1	22.3	19.8	19632	0	6721	3.1
Benson ND	31.4	4.9	-3.3	6964	50.8	49.2	36.1	13.5	13846	-8.4	2149	7.9
Big Horn MT	29.8	2.5	11.8	12671	36.6	60.6	35.8	7.8	13386	11.7	4480	14.4
Big Horn WY	38.7	3.6	8.9	11461	94	6	28.7	16.8	19738	20.2	4566	5.6
Billings ND	41.9	0.8	-19.9	888	98.8	1.2	24.9	16	14166	37.5	1244	3.8
Bottineau ND	43.4	4.2	-10.8	7149	97.2	2.8	22.2	21.3	17261	-23.2	7722	4
Bowman ND	43	2.8	-9.8	3242	99	1	24.1	21.8	22029	10.5	13130	1.7
Burke ND	47.5	2	-25.3	2242	99.2	0.8	20.8	25.1	22074	-4.5	6212	2.8
Butte SD	38	4	14.9	9094	95.5	4.5	28.3	15.2	17926	15.8	7467	3.1
Campbell SD	41.9	2.3	-9.3	1782	99.3	0.7	26.4	22.1	23139	25	0	7.1
Campbell WY	32.2	7	14.7	33698	96.1	3.9	31	5.3	26112	16.6	8739	3.4
Carbon MT	41.9	4.6	18.2	9552	97.1	1.9	24	15.5	20889	17.4	3467	5.1
Carbon WY	38.9	2	-6.1	15639	90.1	9.9	24.1	12.3	22613	14.9	8934	4.1
Carter MT	41.8	0.4	-9.5	1360	98.6	1.4	26.7	15.7	17261	46.4	1707	2.1
Cavalier ND	45.2	3.2	19.7	4831	98.1	1.9	24.6	22.9	27292	17	9526	3.2
Converse WY	37.5	2.8	8.3	12052	94.7	5.3	28.5	11	21453	18.2	5208	4.4
Corson SD	28.3	1.7	-0.3	4181	37.2	62.8	36.9	10.5	14297	26	0	6.5
Crook WY	40.2	2.1	11.2	5887	97.9	2.1	26.9	14.7	22714	14	4412	4.1
Custer SD	43.2	4.7	17.7	7275	94.2	5.8	24.1	16	19739	14.2	0	3.1
Dickey ND	40.7	5	-5.7	5757	97.8	2.2	23.8	21.3	19625	1.7	8308	1.6
Divide ND	49	1.8	-21.2	2283	99	1	20.2	29.5	21879	-6.7	3705	2.2
Dunn ND	40.9	1.7	-10.1	3600	86.6	13.4	27.4	17.4	14444	23.9	8069	3.5
Eddy ND	43.8	4.3	-6.6	2757	96.4	3.6	23.6	24.7	18106	3.6	3461	5
Emmons ND	44.5	2.8	-10.3	4331	99.1	0.9	24.8	25.6	17601	8	8003	4.4
Fall River SD	45.5	4.2	1.4	7453	90.5	9.5	22.9	22.5	22830	24	0	3.8
Fallon MT	41.1	1.7	-8.6	2837	98.6	1.6	25.5	16.3	21607	23.2	7515	3.6
Foster ND	40.5	5.8	-5.6	3759	99	1	26.2	21.4	22452	4.8	10276	2.7
Fremont WY	37.7	3.8	6.4	35804	76.5	23.5	27.5	13.3	20808	17.5	8254	6.4
GoldenValley ND	41.2	1.9	-8.7	1924	97.8	2.2	28.3	21.3	16089	3.7	9281	3.5
Goshen WY	40	5.6	1.3	12538	93.8	6.2	24.2	17.3	21565	23.2	5350	3
Grant ND	46.5	1.7	-19.9	2841	96.9	3.1	23.4	24.7	13774	23	4611	2.8
Griggs ND	45.8	3.8	-16.6	2754	99.3	0.7	22.6	25.7	21306	6.1	7901	2
Harding SD	37.6	0.5	-18.9	1353	97.6	2.4	32.5	13.4	14781	52.8	3292	2.8

County/State	pctwohs	pctcol	pctbepov	pchnfest	dstnmsa	preshied	intaccs	pctsrrt	pctmanf	urban
Adams ND	42.5	7.7	14.1	-4.5	143	0	0	0.25	0.00	0.50
Albany WY	28.3	21.6	13.6	38.2	49.3	1	1	0.24	0.03	0.85
Banner NE	34.6	7.9	12.3	0	91.6	0	0	0.00	0.00	0.09
Barnes ND	40.8	10.1	14.3	9.6	60	1	1	0.20	0.00	0.58
Benson ND	57.2	5.3	29.4	18.6	110	0	0	0.06	0.00	0.14
Big Horn MT	55.3	7	28.8	22.9	46.4	1	1	0.13	0.00	0.27
Big Horn WY	41.5	9.5	12.4	47.8	132.2	0	0	0.11	0.00	0.20
Billings ND	48.1	7.8	16	66.3	132	0	1	0.22	0.00	0.11
Bottineau ND	40.5	9.5	14	21.5	192	1	0	0.16	0.00	0.33
Bowman ND	41.2	9.3	13.6	9.4	192	0	0	0.24	0.00	0.49
Burke ND	46	6.2	15.8	-29.5	228	0	0	0.18	0.00	0.18
Butte SD	44.3	7.9	17.1	25	60.1	0	0	0.18	0.00	0.50
Campbell SD	47.3	7.4	16.7	32.5	105.8	0	0	0.16	0.00	0.27
Campbell WY	41.8	9	7.6	23	140.9	1	1	0.19	0.00	0.58
Carbon MT	32.5	13.4	13.4	43.2	61.1	0	0	0.17	0.00	0.23
Carbon WY	39.7	8.9	11.8	15.9	121	0	1	0.23	0.00	0.55
Carter MT	40.5	7.4	19.2	-15.9	169.8	0	0	0.05	0.00	0.30
Cavalier ND	44.4	8.7	14	27.4	106	0	0	0.17	0.00	0.43
Converse WY	41.7	7.7	12.4	20	50.4	0	1	0.15	0.00	0.44
Corson SD	59.3	5.8	35.3	0.4	87.7	0	0	0.08	0.00	0.19
Crook WY	40.3	9.8	9.7	27.9	79.6	0	1	0.10	0.00	0.20
Custer SD	34.4	11.7	12.5	0.2	41.6	0	0	0.15	0.00	0.26
Dickey ND	44.7	10.4	15.9	28.6	141	0	0	0.19	0.00	0.34
Divide ND	40.5	9.3	14.6	8.1	178	0	0	0.13	0.00	0.48
Dunn ND	48.9	6.4	18.3	48.5	122	0	0	0.09	0.00	0.20
Eddy ND	45.7	7.7	14.3	-14.1	120	0	0	0.12	0.00	0.53
Emmons ND	54.5	6.2	16.2	79.8	66	0	0	0.13	0.00	0.31
Fall River SD	34.8	11.8	16.2	54.1	72.8	0	0	0.25	0.00	0.55
Fallon MT	44.3	6.9	13.6	33.3	221.1	0	0	0.14	0.00	0.60
Foster ND	45.5	8.1	12.8	57.2	123	0	0	0.20	0.00	0.60
Fremont WY	42.3	10.1	17	26.8	148.6	0	0	0.25	0.03	0.26
GoldenValley ND	42.9	9.9	18.4	4.2	158	0	1	0.20	0.00	0.58
Goshen WY	42	9.3	16	25.1	83.6	1	0	0.15	0.00	0.46
Grant ND	50.5	6.2	25	31.2	64	0	0	0.09	0.00	0.23
Griggs ND	44.6	8.4	15	69.5	85	0	0	0.15	0.00	0.38
Harding SD	39.8	9.8	15.4	-16.7	130.6	0	0	0.15	0.00	0.28

County/State	mage2000	psm2000	popc9000	tpop00	pctwhite	pctnonwh	pct17und	pct65ovr	pci1999	pciprch	rtsapc99	avgunp00
Hettinger ND	46.2	2.4	-21.2	2715	98.9	1.1	23.4	25.2	20892	-7.4	5701	2.6
Hot Springs WY	44.2	2.5	1.5	4882	96	4	22	20	23392	10.1	4327	3.4
Johnson WY	43	1.7	15.1	7075	97	3	24.2	18	23652	18.1	5694	2.8
Kidder ND	44.5	1.9	-17.4	2753	99.5	0.5	23.2	24	17290	13.7	4751	4.8
Kimball NE	42.8	4.3	-0.5	4089	97	3	24.7	19.1	22224	14	5241	2.3
Kittson MN	42.4	4.8	-8.4	5285	98.1	1.9	25.1	18.9	25843	18.5	7091	7.8
La Moure ND	43.3	4.1	-12.7	4701	99.2	0.8	24.2	23.4	18444	-11.1	3821	2.5
Lincoln WY	36.8	3.6	15.4	14573	97.1	2.9	30.9	12.4	20870	20.2	6238	5.3
Logan ND	46.4	2.3	-5.8	2308	99.2	0.8	22.6	27	19969	5.3	4908	1.8
Marshall MN	40.5	5.6	-7.6	10155	97.2	2.8	25.4	16.9	20887	1.8	6846	11.3
Marshall SD	41.6	3.3	-5.5	2904	99.3	0.7	42.6	33.5	26659	22.1	0	1.8
McHenry ND	43	3.2	2	5987	98.7	1.3	24	21.8	15414	-3	2316	5.1
McIntosh ND	51	3.4	-15.7	3390	98.9	1.1	19.4	34.2	21188	13.2	8150	2.4
McKenzie ND	39.5	2	-10.1	5737	77.4	22.6	30.6	15.7	19955	13.8	4139	3.1
McLean ND	44.1	4	-11	9311	92.5	7.5	23.8	20.4	20229	-4.5	4819	6
McPherson SD	47.6	4	-10	4576	92.6	7.4	14.1	18.8	20999	18.6	0	4.1
Mercer ND	40.1	7.7	-11.9	8644	96	4	29.1	14.3	23021	3.2	5919	5.6
Mountrail ND	39.6	3.4	-5.6	6631	66	34	28.1	17.7	20139	-1.8	7041	5.5
Nelson ND	47.2	3.7	-15.8	3715	98.6	1.4	22.1	27.4	21059	5.4	4036	4.6
Niobrara WY	42.8	0.9	-3.7	2407	98	2	22.6	18.7	21662	23.3	4512	2.7
Norman MN	40.9	8.5	-6.7	7442	95.3	4.7	25.7	18.7	23327	8	10523	5.9
Oliver ND	42	2.8	-13.3	2065	97.6	2.4	27.4	14.2	17778	9.3	1796	6
Park MT	40.6	5.9	8.1	15694	96.6	3.4	23.5	13.6	19514	11	6395	5.3
Park WY	39.8	3.7	11.3	25786	96.5	3.5	24.4	14.5	25965	17.8	10571	4.1
Pembina ND	41.6	7.7	-7.1	8585	95.5	4.5	24.9	19.5	29339	12	11923	6.9
Perkins SD	43.1	1.2	-14.5	3363	96.6	3.4	24.1	23.7	21987	20.5	4454	2.2
Pierce ND	42.9	4.4	-7.5	4675	98.5	1.5	23.9	24.1	19284	2.6	9193	2.5
Platte WY	41.2	4.1	8.1	8807	96.2	3.8	25.4	16.6	22654	25.4	8010	3.6
Powder River MT	42.1	0.6	-11.1	1858	97.4	2.6	26.6	16.8	18683	37.4	4735	3
Ramsey ND	39.5	9.2	-4.8	12066	92.3	7.7	25	18.8	22878	7.5	11440	3.5
Ransom ND	40.7	6.8	-0.5	5890	97.9	2.1	25	21.2	22398	3.2	6775	2.2
Renville ND	43.6	3	-17.4	2610	97.7	2.3	23.3	22	16533	-33.5	9721	2.1
Richland MT	39.2	4.6	-9.8	9667	96.6	3.4	27.5	14.4	20202	13	8215	6.2
Richland ND	35.4	12.4	-0.8	17998	96.8	3.2	24.7	15.3	23091	12.6	7586	2.6
Roberts SD	37.1	8.9	1	10016	68.3	31.7	30	17	18510	6.3	0	4
Rolette ND	28.9	14.5	7.1	13674	25.1	74.9	36.5	9.7	14916	6.9	5636	12.4

County/State	mage2000	psm2000	popc9000	tpop00	pctwhite	pctnonwh	pct17und	pct65ovr	pci1999	pciprch	rtsapc99	avgunp00
Roosevelt MT	32.3	4.4	-3.4	10620	40.9	59.1	34.6	10.6	16678	13.4	4768	9.5
Sargent ND	40.3	5	-4	4366	98.2	1.8	26.5	16.9	28435	13.5	5753	1.8
Sheridan MT	45.1	2.4	-13.3	4105	97	3	22.9	21	24284	23.3	6086	4.4
Sheridan ND	48.1	1.7	-20.4	1710	99.2	0.8	21.4	26.6	18276	2.9	3450	6.1
Sheridan WY	40.6	10.4	12.7	26560	95.9	4.1	24.1	15.5	28344	16	9575	4.2
Sioux ND	23.9	3.5	7.5	4044	14.3	85.7	40.3	5.6	11023	11.7	3405	6.2
Sioux NE	41.5	0.7	-4.8	1475	97.6	2.4	24.3	15.3	11147	-0.4	1425	1.2
Slope ND	42.5	0.6	-15.4	767	99.7	0.3	25.3	17.9	12097	-9.8	0	2.8
Steele ND	41.4	3.2	-6.7	2258	98.3	1.7	27.3	19.6	20440	-8	5434	1.4
Stutsman ND	39.6	9.5	-1.5	21908	97.5	2.5	22.9	17.6	23434	9.6	9864	2.2
Sublette WY	39.8	1.2	22.2	5920	97.5	2.5	25.8	12	24992	20.1	4676	2.5
Sweetwater WY	34.2	3.6	-3.1	37613	91.6	8.4	28.9	8	26347	15	10142	4.9
Teton WY	35	4.3	63.3	18251	93.6	6.4	19.9	6.9	59632	25.8	21662	1.7
Towner ND	44	2.7	-20.7	2876	97.3	2.7	24.6	23.3	18602	-7.6	3384	2.9
Traill ND	39	9.8	-3.1	8477	97.3	2.7	24.8	19.1	22095	4.6	5703	3
Traverse MN	42.9	7.1	-7.4	4134	96.4	3.6	25.3	22.7	22987	-0.6	7400	5.9
Uinta WY	31.4	9.5	5.5	19742	94.3	5.7	33.5	7	20486	13.7	8591	5.5
Walsh ND	40.9	9.5	-10.5	12389	94.9	5.1	25	19.3	21498	6	8477	4.2
Washakie WY	39.4	3.7	-1.2	8289	90.2	9.8	27.2	15.9	23795	18.7	9375	5.3
Wells ND	45.2	3.9	-13	5102	99.1	0.9	22.5	26	21766	-5.2	7923	3.7
Weston WY	40.7	2.8	1.9	6644	95.9	2.5	24.1	15.6	24872	20.9	5898	4.5
Wibaux MT	42.3	1.2	-10.3	1068	98	2	25.8	18.7	16793	20.7	0	4.1
Wilkin MN	38.1	9.6	-5	7138	97.8	2.2	27.8	14	22478	-2.5	5167	2.7
Williams ND	39.8	9.1	-6.5	19761	92.9	7.1	26.2	16.5	21736	6.2	10219	3.7

County/State	pctwohs	pctcol	pctbepov	pchnfest	dstnmsa	preshied	intaccs	pctsrrt	pctmanf	urban
Roosevelt MT	53.4	6.5	31.7	-3.2	214.4	1	0	0.19	0.00	0.25
Sargent ND	44.8	6.5	9.6	26	93	0	0	0.11	0.04	0.16
Sheridan MT	40.1	8.1	13.7	6.1	117.5	0	0	0.17	0.00	0.50
Sheridan ND	58.5	5.9	23.1	8.3	64	0	0	0.13	0.00	0.24
Sheridan WY	34.1	11.7	11.5	33.9	129.1	1	1	0.25	0.00	0.60
Sioux ND	62.7	4.7	35.3	207.2	67	0	0	0.06	0.00	0.21
Sioux NE	34.9	12.2	11.8	-82.1	188	0	0	0.05	0.00	0.19
Slope ND	47.2	6.7	17.4	-18.5	189	0	0	0.04	0.00	0.18
Steele ND	40.6	9.5	12.9	20.1	66	1	0	0.08	0.00	0.23
Stutsman ND	41.1	10.9	12.7	26.2	93	1	1	0.26	0.13	0.71
Sublette WY	30.5	14.1	8.4	21.9	237.1	0	0	0.13	0.00	0.24
Sweetwater WY	44.9	7.7	8.1	12.3	169.4	1	1	0.25	0.03	0.50
Teton WY	16.7	20.5	4.9	70.7	295	0	0	0.53	0.00	0.47
Towner ND	42.8	8.6	16.4	-5.7	123	0	0	0.23	0.00	0.47
Traill ND	39.1	11.4	11.1	11.2	40	1	1	0.22	0.00	0.23
Traverse MN	44	6.8	14.7	17.4	80	0	0	0.20	0.00	0.39
Uinta WY	47.7	7.6	10.4	18	81.2	0	1	0.20	0.00	0.58
Walsh ND	46.6	8.6	13.3	12.3	39	0	1	0.19	0.00	0.36
Washakie WY	37	11.9	10.7	27.3	167.1	0	0	0.23	0.00	0.63
Wells ND	46.9	8.1	14.6	16.2	112	0	0	0.18	0.00	0.39
Weston WY	38.6	8.1	11.1	-3.8	78.6	0	0	0.15	0.00	0.46
Wibaux MT	46.9	7.3	18.3	80.2	167.3	0	1	0.10	0.00	0.53
Wilkin MN	45.2	7.4	11.1	19.2	45.9	0	0	0.12	0.00	0.50
Williams ND	41.7	9.2	14.2	5.4	225	1	0	0.30	0.00	0.63

Appendix B

Factor Analysis Procedures and Results

The FACTOR Procedure
Initial Factor Method: Principal Components

Prior Communality Estimates: ONE

Eigenvalues of the Correlation Matrix: Total = 22 Average = 1

	Eigenvalue	Difference	Proportion	Cumulative
1	5.62882989	0.62520085	0.2559	0.2559
2	5.00362905	3.01185770	0.2274	0.4833
3	1.99177135	0.51180926	0.0905	0.5738
4	1.47996208	0.26707395	0.0673	0.6411
5	1.21288814	0.20585082	0.0551	0.6962
6	1.00703731	0.16538557	0.0458	0.7420
7	0.84165174	0.04680180	0.0383	0.7803
8	0.79484994	0.13194698	0.0361	0.8164
9	0.66290296	0.06673651	0.0301	0.8465
10	0.59616645	0.10022635	0.0271	0.8736
11	0.49594010	0.04347987	0.0225	0.8962
12	0.45246023	0.08449747	0.0206	0.9167
13	0.36796277	0.04182593	0.0167	0.9335
14	0.32613684	0.01932398	0.0148	0.9483
15	0.30681286	0.03138544	0.0139	0.9622
16	0.27542742	0.07509669	0.0125	0.9747
17	0.20033073	0.07263205	0.0091	0.9839
18	0.12769868	0.00572717	0.0058	0.9897
19	0.12197151	0.04965167	0.0055	0.9952
20	0.07231984	0.03920755	0.0033	0.9985
21	0.03311229	0.03297447	0.0015	1.0000
22	0.00013782		0.0000	1.0000

4 factors will be retained by the NFACTOR criterion.

The FACTOR Procedure
Initial Factor Method: Principal Components

Scree Plot of Eigenvalues

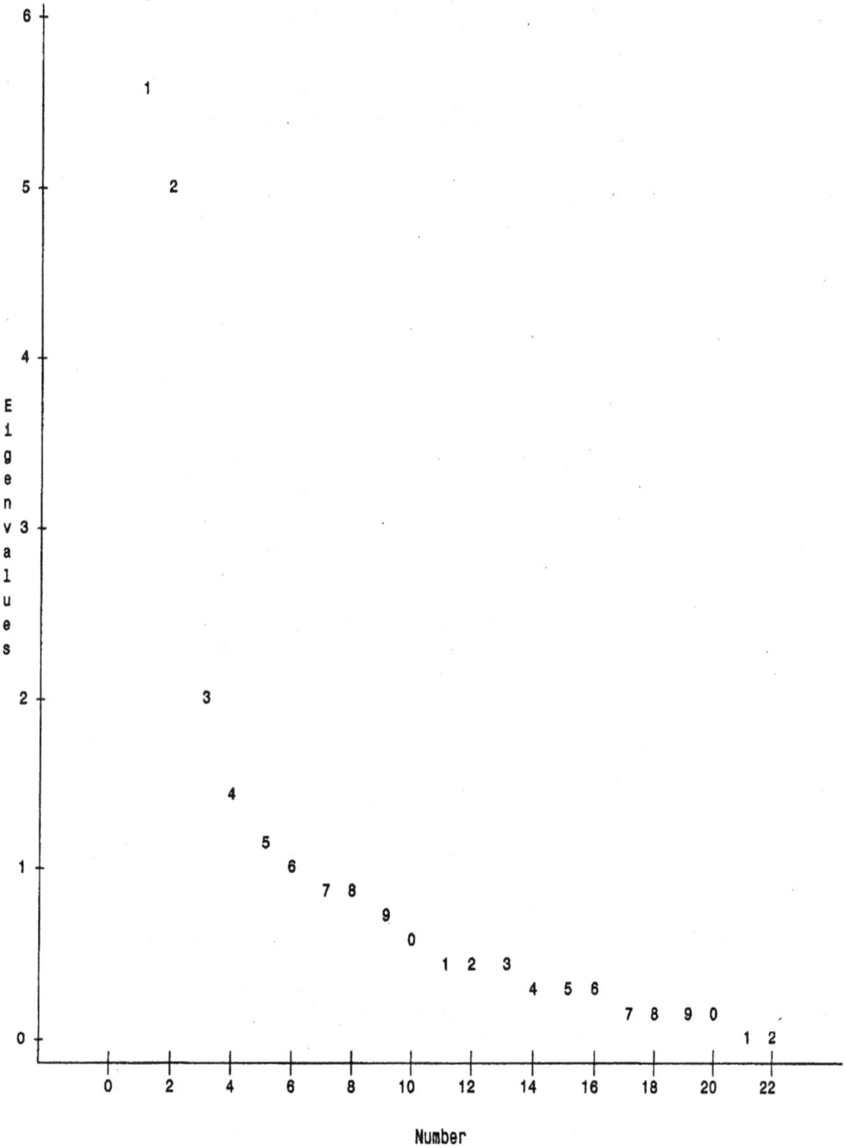

The FACTOR Procedure
Initial Factor Method: Principal Components

Factor Pattern

	Factor1	Factor2	Factor3	Factor4
pctwhite	0.74895	-0.54853	0.13387	-0.20853
pci1999	0.71656	0.30317	-0.21429	0.26133
pctcol	0.69421	0.35334	-0.28599	-0.10572
pctsrrt	0.61593	0.38691	-0.04474	0.40941
rtsapc99	0.56372	0.37361	0.08998	0.47024
urban	0.55267	0.23707	0.14646	0.12857
avgunp00	-0.58295	0.38407	0.07174	0.10683
pct17und	-0.63377	0.44316	-0.16281	-0.04661
pctnonwh	-0.74897	0.54733	-0.13398	0.21161
pctwohs	-0.79944	-0.12615	0.31670	0.19432
pctbepov	-0.83967	0.10197	-0.07757	0.18012
tpop00	0.30984	0.76367	0.23008	-0.05945
popc9000	0.27598	0.67001	-0.38834	-0.08753
preshied	0.09373	0.57043	0.45362	0.04374
pchnfest	-0.08462	0.37684	-0.12488	0.16966
pct65ovr	0.12651	-0.78902	0.23535	0.32799
mage2000	0.33189	-0.86176	0.08775	0.12478
psm2000	0.12040	0.53759	0.61336	0.18630
pctmanf	0.17814	0.29613	0.43959	-0.27574
pciprch	0.07788	0.23896	-0.47995	-0.45880
dstnmsa	0.16491	0.01914	-0.48534	0.37016
intaccs	0.17197	0.41455	0.36756	-0.44707

Variance Explained by Each Factor

Factor1	Factor2	Factor3	Factor4
5.6288299	5.0036290	1.9917713	1.4799621

Final Communality Estimates: Total = 14.104192

mage2000	urban	pctsrrt	popc9000	tpop00	pctwhite	pctnonwh	pct17und
0.87605388	0.39962203	0.69868813	0.68354837	0.73567414	0.92322136	0.92325206	0.62673766

pct65ovr	pci1999	rtsapc99	pctwohs	pctcol	pctbepov	preshied	dstnmsa
0.80152112	0.71958887	0.68658454	0.79307797	0.69974048	0.75389990	0.54185372	0.40013941

intaccs	pciprch	pctmanf	pchnfest	avgunp00	psm2000
0.53639881	0.50401877	0.38870667	0.19354377	0.50389758	0.71442313

The FACTOR Procedure
Rotation Method: Varimax

Orthogonal Transformation Matrix

	1	2	3	4
1	-0.71454	0.65285	0.13933	0.20931
2	0.62787	0.45762	0.49898	0.38390
3	-0.13870	-0.11348	0.75961	-0.62520
4	0.27564	0.59286	-0.39319	-0.64648

Rotated Factor Pattern

	Factor1	Factor2	Factor3	Factor4
pctnonwh	0.95573	-0.09783	-0.01622	0.00032
pct17und	0.74083	-0.22012	0.02747	0.16940
pctbepov	0.72441	-0.38592	-0.19585	-0.20455
avgunp00	0.67718	-0.14962	0.12291	-0.08849
pchnfest	0.36115	0.23196	0.01468	0.09535
mage2000	-0.75600	-0.11366	-0.36616	-0.39689
pctwhite	-0.95561	0.09911	0.01433	-0.00271
pctsrrt	-0.07812	0.82697	0.08392	0.04075
rtsapc99	-0.05108	0.80757	0.14843	-0.09884
pci1999	-0.21990	0.78580	-0.01441	0.23140
pctcol	-0.26365	0.58469	0.09736	0.52810
urban	-0.23093	0.52890	0.25600	0.03200
intaccs	-0.03681	-0.00478	0.68580	0.25436
psm2000	0.21779	0.36546	0.67769	-0.27233
preshied	0.24032	0.29669	0.62507	-0.07327
tpop00	0.20980	0.49040	0.62238	0.25261
pctmanf	-0.07833	0.03845	0.61492	0.05440
dstnmsa	0.06353	0.39095	-0.48169	0.10600
pciprch	0.03449	-0.05734	-0.05410	0.70471
popc9000	0.25322	0.47896	0.11220	0.61436
pctwohs	0.50166	-0.50038	-0.01017	-0.53938
pct65ovr	-0.52803	-0.11073	-0.32627	-0.63560

Variance Explained by Each Factor

Factor1	Factor2	Factor3	Factor4
4.9971700	3.9927845	2.7331473	2.3810906

The FACTOR Procedure
Rotation Method: Varimax

Final Communality Estimates: Total = 14.104192

mage2000	urban	pctsrrt	popc9000	tpop00	pctwhite	pctnonwh	pct17und
0.87605388	0.39962203	0.69868813	0.68354837	0.73567414	0.92322136	0.92325206	0.62673766

pct65ovr	pci1999	rtsapc99	pctwohs	pctcol	pctbepov	preshied	dstnmsa
0.80152112	0.71958887	0.68658454	0.79307797	0.69974048	0.75389990	0.54185372	0.40013941

intaccs	pciprch	pctmanf	pchnfest	avgunp00	psm2000
0.53639881	0.50401877	0.38870667	0.19354377	0.50389758	0.71442313

The FACTOR Procedure
Rotation Method: Varimax

Scoring Coefficients Estimated by Regression

Squared Multiple Correlations of the Variables with Each Factor

Factor1	Factor2	Factor3	Factor4
1.0000000	1.0000000	1.0000000	1.0000000

Standardized Scoring Coefficients

	Factor1	Factor2	Factor3	Factor4
pctnonwh	0.21250	0.05559	-0.07127	-0.03624
pct17und	0.13872	-0.04237	-0.02120	0.08190
pctbepov	0.15833	-0.01149	-0.08805	-0.07773
avgunp00	0.13710	0.00622	0.02285	-0.06140
pchnfest	0.09832	0.09973	-0.05721	-0.00915
mage2000	-0.13314	0.00467	-0.07741	-0.13583
pctwhite	-0.21207	-0.05447	0.07029	0.03483
pctsrrt	0.04973	0.27338	-0.07200	-0.11221
rtsapc99	0.05664	0.28280	-0.03940	-0.18403
pci1999	0.01068	0.22773	-0.10318	0.00302
pctcol	-0.04356	0.08678	-0.02856	0.18887
urban	-0.02666	0.12894	0.05902	-0.06340
intaccs	-0.07867	-0.14217	0.30455	0.11811
psm2000	0.04416	0.10282	0.24102	-0.22818
preshied	0.03624	0.05472	0.22058	-0.11424
tpop00	0.02940	0.06886	0.18737	0.02386
pctmanf	-0.06742	-0.08776	0.27485	0.01181
dstnmsa	0.08421	0.19681	-0.27745	-0.00175
pciprch	-0.03193	-0.12556	-0.03539	0.37230
popc9000	0.05978	0.08035	-0.05120	0.22180
pctwohs	0.09979	-0.04446	0.03679	-0.22370
pct65ovr	-0.07037	0.06050	-0.07294	-0.27298

The FACTOR Procedure
Initial Factor Method: Principal Components

Prior Communality Estimates: ONE

Eigenvalues of the Correlation Matrix: Total = 17 Average = 1

	Eigenvalue	Difference	Proportion	Cumulative
1	5.28929448	0.91055161	0.3111	0.3111
2	4.37874287	2.87991810	0.2576	0.5687
3	1.49882477	0.23240253	0.0882	0.6569
4	1.26642224	0.38203066	0.0745	0.7314
5	0.88439159	0.18181387	0.0520	0.7834
6	0.70257772	0.00653271	0.0413	0.8247
7	0.69604501	0.15664062	0.0409	0.8657
8	0.53940439	0.05420172	0.0317	0.8974
9	0.48520267	0.14487667	0.0285	0.9259
10	0.34032600	0.04242489	0.0200	0.9460
11	0.29790111	0.08169923	0.0175	0.9635
12	0.21620189	0.08151345	0.0127	0.9762
13	0.13468844	0.00245785	0.0079	0.9841
14	0.13223059	0.03586769	0.0078	0.9919
15	0.09636290	0.05516231	0.0057	0.9976
16	0.04120059	0.04101786	0.0024	1.0000
17	0.00018273		0.0000	1.0000

4 factors will be retained by the NFACTOR criterion.

The FACTOR Procedure
Initial Factor Method: Principal Components

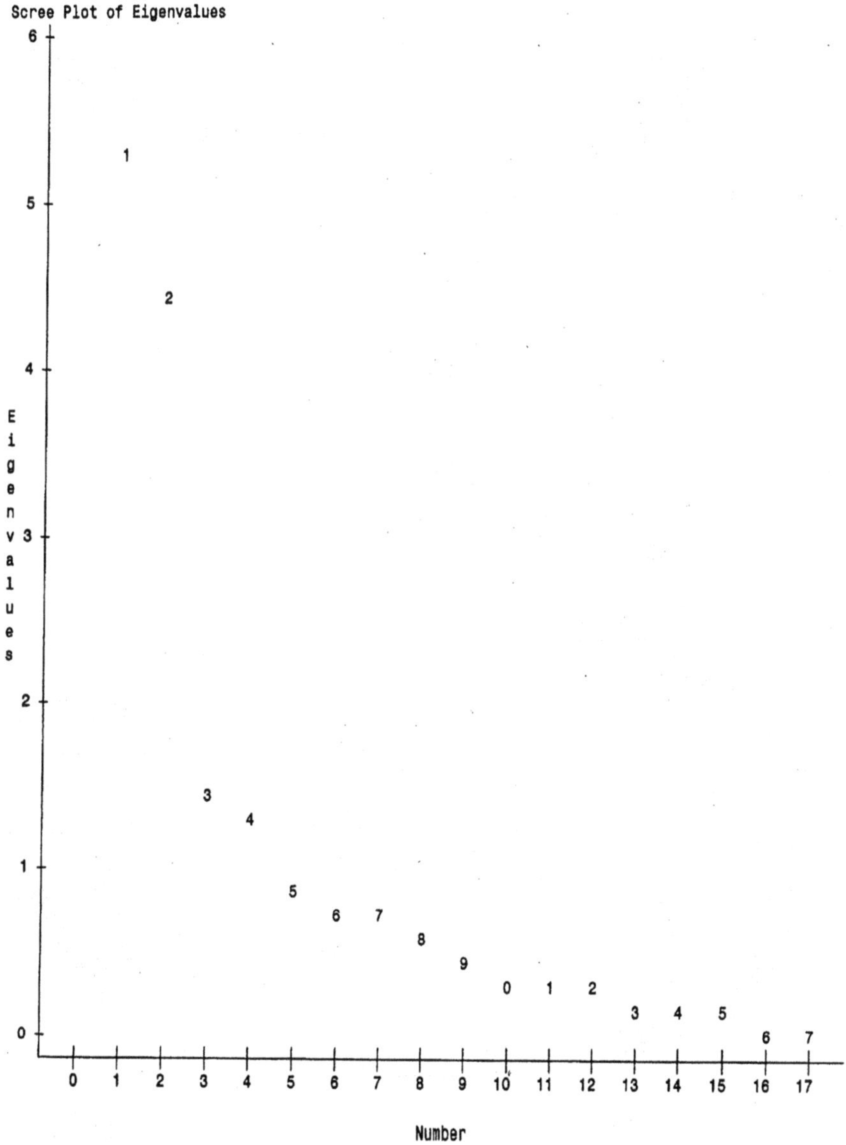

The FACTOR Procedure
Initial Factor Method: Principal Components

Factor Pattern

	Factor1	Factor2	Factor3	Factor4
pci1999	0.76445	0.19801	0.30132	0.04744
pctcol	0.73655	0.30523	0.13695	-0.22831
pctsrrt	0.67155	0.28844	0.15367	0.41261
pctwhite	0.66742	-0.65234	-0.19196	-0.13250
rtsapc99	0.61838	0.28605	0.10212	0.41365
urban	0.56427	0.16500	-0.20499	0.40693
pct17und	-0.57801	0.53176	0.08977	-0.06310
pctnonwh	-0.66818	0.65118	0.19481	0.13507
pctwohs	-0.81408	-0.07160	-0.14835	0.34780
pctbepov	-0.82098	0.22110	0.11874	0.26406
tpop00	0.40107	0.70338	-0.28395	0.09622
popc9000	0.36859	0.64885	0.26520	-0.37178
preshied	0.14754	0.53969	-0.39881	0.32753
pct65ovr	0.02440	-0.82901	0.00706	0.37122
mage2000	0.23136	-0.91238	0.03683	0.11379
dstnmsa	0.17660	0.08324	0.64733	0.24902
intaccs	0.21004	0.36624	-0.68102	-0.08432

Variance Explained by Each Factor

Factor1	Factor2	Factor3	Factor4
5.2892945	4.3787429	1.4988248	1.2664222

Final Communality Estimates: Total = 12.433284

mage2000	urban	pctsrrt	popc9000	tpop00	pctwhite
0.90026038	0.55323659	0.72803791	0.76541959	0.74548312	0.92541063

pctnonwh	pct17und	pct65ovr	pci1999	rtsapc99	pctwohs
0.92668801	0.62890234	0.82570825	0.71663108	0.64575289	0.81082287

pctcol	pctbepov	preshied	dstnmsa	intaccs
0.70654555	0.80671639	0.57935582	0.51916855	0.64914438

The FACTOR Procedure
Rotation Method: Varimax

Orthogonal Transformation Matrix

	1	2	3	4
1	-0.58579	0.57572	0.56773	0.05548
2	0.78440	0.42443	0.35108	0.28515
3	0.18804	0.28947	-0.00782	-0.93850
4	0.07877	-0.63609	0.74455	-0.18661

Rotated Factor Pattern

	Factor1	Factor2	Factor3	Factor4
pctnonwh	0.94947	-0.13783	-0.05168	-0.05942
pct17und	0.76761	-0.04095	-0.18915	0.04710
pctbepov	0.69748	-0.51241	-0.19280	-0.14322
pct65ovr	-0.63400	-0.57190	-0.00086	-0.31094
mage2000	-0.83531	-0.31576	-0.10454	-0.30313
pctwhite	-0.94920	0.13609	0.05274	0.05589
popc9000	0.31363	0.80085	0.15818	0.02596
pctcol	-0.18427	0.73846	0.35426	0.04198
pci1999	-0.23208	0.58120	0.53649	-0.19277
pctwohs	0.42021	-0.76325	-0.22720	0.00875
pctsrrt	-0.10574	0.29108	0.78853	-0.10171
rtsapc99	-0.08607	0.24387	0.75869	-0.05716
urban	-0.20761	0.07671	0.68286	0.19480
tpop00	0.27098	0.38604	0.54850	0.47135
preshied	0.28771	-0.00978	0.52022	0.47524
intaccs	0.02954	0.13287	0.19036	0.77096
dstnmsa	0.10318	0.16598	0.30984	-0.62046

Variance Explained by Each Factor

Factor1	Factor2	Factor3	Factor4
4.5700514	3.1799619	2.9467024	1.7365686

Final Communality Estimates: Total = 12.433284

mage2000	urban	pctsrrt	popc9000	tpop00	pctwhite
0.90026038	0.55323659	0.72803791	0.76541959	0.74548312	0.92541063

The FACTOR Procedure
Rotation Method: Varimax

pctnonwh	pct17und	pct65ovr	pci1999	rtsapc99	pctwohs
0.92668801	0.62890234	0.82570825	0.71663108	0.64575289	0.81082287

pctcol	pctbepov	preshied	dstnmsa	intaccs
0.70654555	0.80671639	0.57935582	0.51916855	0.64914438

The FACTOR Procedure
Rotation Method: Varimax

Scoring Coefficients Estimated by Regression

Squared Multiple Correlations of the Variables with Each Factor

Factor1	Factor2	Factor3	Factor4
1.0000000	1.0000000	1.0000000	1.0000000

Standardized Scoring Coefficients

	Factor1	Factor2	Factor3	Factor4
pctnonwh	0.22349	-0.03983	0.05889	-0.10649
pct17und	0.16661	0.03766	-0.05697	-0.01834
pctbepov	0.16185	-0.17763	0.08423	-0.10748
pct65ovr	-0.12723	-0.26279	0.15436	-0.11285
mage2000	-0.17737	-0.11329	0.01839	-0.09682
pctwhite	-0.22310	0.03889	-0.05756	0.10424
popc9000	0.08556	0.34097	-0.12837	-0.06515
pctcol	-0.02391	0.25088	-0.03141	-0.02451
pci1999	-0.00844	0.13677	0.12425	-0.17475
pctwohs	0.08035	-0.29889	0.11213	0.02844
pctsrrt	0.02224	-0.07651	0.33699	-0.13120
rtsapc99	0.02130	-0.09301	0.33197	-0.09978
urban	-0.03334	-0.16657	0.31411	0.08506
tpop00	0.05195	0.00866	0.15750	0.21363
preshied	0.05068	-0.17316	0.25375	0.23815
intaccs	-0.04834	-0.03081	0.00589	0.46490
dstnmsa	0.09206	0.02723	0.16866	-0.43475

Appendix C

Factor and Broadband Development Index Scores by County

County/State	Factor1	Factor2	Factor3	Factor4	AF1	AF2	AF3	AF4	Index
Teton WY	0.41	5.27	3.11	-3.73	-0.107	1.204	0.294	-0.26	1.131
Albany WY	0.20	1.80	1.95	2.67	-0.053	0.411	0.184	0.19	0.730
Park MT	-0.34	1.49	-0.20	0.95	0.088	0.340	-0.019	0.07	0.475
Sheridan WY	-0.10	0.76	1.77	1.40	0.026	0.174	0.168	0.10	0.466
Sublette WY	-0.02	2.77	-1.06	-1.39	0.006	0.633	-0.101	-0.10	0.441
Johnson WY	-0.55	1.18	-0.19	0.62	0.140	0.271	-0.018	0.04	0.436
Carbon MT	-0.36	1.95	-1.32	0.04	0.093	0.445	-0.125	0.00	0.415
Custer SD	-0.39	1.69	-1.71	0.27	0.101	0.387	-0.162	0.02	0.344
Crook WY	-0.34	1.38	-1.46	1.11	0.086	0.315	-0.138	0.08	0.341
Campbell WY	0.75	1.15	1.20	2.16	-0.192	0.264	0.114	0.15	0.337
Park WY	0.12	0.99	1.14	0.38	-0.030	0.227	0.108	0.03	0.332
Traill ND	-0.42	0.28	0.10	1.88	0.107	0.065	0.010	0.13	0.313
Stutsman ND	-0.28	-0.38	2.09	1.83	0.072	-0.086	0.198	0.13	0.312
Platte WY	-0.31	0.42	-0.15	1.03	0.079	0.096	-0.014	0.07	0.234
Barnes ND	-0.51	-0.59	1.02	2.01	0.130	-0.135	0.097	0.14	0.233
Converse WY	0.06	0.83	-0.68	1.63	-0.014	0.189	-0.064	0.11	0.225
Pembina ND	-0.44	0.20	0.36	0.48	0.113	0.045	0.034	0.03	0.225
Weston WY	-0.39	0.67	-0.35	-0.01	0.099	0.153	-0.033	0.00	0.218
Carbon WY	-0.05	0.28	0.78	0.91	0.012	0.065	0.074	0.06	0.214
Fall River SD	-0.55	0.43	-0.08	-0.30	0.140	0.099	-0.007	-0.02	0.211
Hot Springs WY	-0.57	0.20	0.61	-0.72	0.146	0.047	0.058	-0.05	0.200
Richland ND	-0.01	-0.23	0.76	2.33	0.004	-0.051	0.072	0.16	0.188
Sweetwater WY	0.72	0.20	2.02	1.80	-0.184	0.047	0.191	0.13	0.180
Washakie WY	0.03	0.72	0.87	-0.87	-0.009	0.165	0.082	-0.06	0.178
Cavalier ND	-0.41	0.44	0.24	-0.85	0.106	0.100	0.023	-0.06	0.169
Sioux NE	-0.36	1.21	-1.86	-0.45	0.093	0.277	-0.176	-0.03	0.162
Kimball NE	-0.52	0.02	0.24	-0.05	0.134	0.004	0.023	0.00	0.158
Ransom ND	-0.51	0.17	-0.14	-0.06	0.132	0.039	-0.013	0.00	0.153
Walsh ND	-0.42	-0.37	0.20	1.40	0.108	-0.084	0.018	0.10	0.140
Uinta WY	0.67	0.65	0.26	1.77	-0.172	0.149	0.025	0.12	0.126
Niobrara WY	-0.51	0.08	-0.02	-0.33	0.132	0.018	-0.002	-0.02	0.126
Lincoln WY	0.39	1.37	-0.82	-0.16	-0.100	0.312	-0.078	-0.01	0.124
Kittson MN	-0.46	0.23	-0.25	-0.33	0.118	0.052	-0.023	-0.02	0.124
Sargent ND	-0.30	0.57	-0.88	-0.29	0.078	0.130	-0.084	-0.02	0.104
Wilkin MN	-0.18	0.26	-0.58	0.59	0.047	0.060	-0.055	0.04	0.094
Sheridan MT	-0.69	-0.30	0.25	-0.53	0.177	-0.070	0.024	-0.04	0.093
Nelson ND	-1.06	-0.58	-0.34	-0.26	0.272	-0.132	-0.032	-0.02	0.090
Big Horn WY	0.17	1.10	-1.08	-0.26	-0.045	0.252	-0.102	-0.02	0.087
Banner NE	-0.20	1.14	-2.61	0.24	0.052	0.261	-0.247	0.02	0.082
Steele ND	-0.31	0.01	-0.52	0.66	0.079	0.003	-0.049	0.05	0.079
Ramsey ND	0.06	-0.53	2.09	0.22	-0.015	-0.122	0.198	0.02	0.077
Goshen WY	0.03	-0.11	0.49	0.70	-0.009	-0.026	0.047	0.05	0.062
Bowman ND	-0.43	-0.37	1.32	-1.38	0.110	-0.084	0.125	-0.10	0.054
Divide ND	-1.14	-0.86	0.29	-1.07	0.292	-0.196	0.027	-0.08	0.048
Mercer ND	-0.18	0.02	-0.31	0.34	0.045	0.005	-0.030	0.02	0.044
Foster ND	-0.35	-0.44	0.86	-0.55	0.090	-0.101	0.082	-0.04	0.032
Dickey ND	-0.26	-0.06	0.22	-0.70	0.068	-0.013	0.021	-0.05	0.027
Butte SD	0.25	0.40	-0.19	0.22	-0.064	0.092	-0.018	0.02	0.025
Griggs ND	-0.86	-0.81	0.15	-0.37	0.221	-0.186	0.014	-0.03	0.024
Marshall MN	-0.27	-0.04	-0.82	0.51	0.069	-0.008	-0.078	0.04	0.018
Hettinger ND	-0.87	-0.67	-0.37	-0.36	0.223	-0.154	-0.035	-0.03	0.010

County/State	Factor1	Factor2	Factor3	Factor4	AF1	AF2	AF3	AF4	Index
Bottineau ND	-0.31	-0.55	0.81	-0.32	0.081	-0.126	0.076	-0.02	0.009
La Moure ND	-0.62	-0.24	-0.91	-0.16	0.160	-0.054	-0.086	-0.01	0.009
Powder River MT	-0.19	0.38	-0.90	-0.64	0.048	0.088	-0.086	-0.05	0.006
GoldenValley ND	-0.22	-0.60	0.71	0.26	0.056	-0.136	0.067	0.02	0.005
Norman MN	-0.21	-0.18	-0.07	-0.06	0.053	-0.042	-0.006	0.00	0.001
Wells ND	-0.73	-0.84	0.45	-0.58	0.189	-0.191	0.042	-0.04	-0.001
Traverse MN	-0.43	-0.48	0.24	-0.39	0.111	-0.110	0.022	-0.03	-0.004
Pierce ND	-0.48	-0.72	0.87	-0.68	0.122	-0.164	0.083	-0.05	-0.007
McLean ND	-0.40	-0.27	-0.69	0.18	0.102	-0.061	-0.065	0.01	-0.012
Adams ND	-0.70	-1.10	1.30	-0.95	0.179	-0.251	0.123	-0.07	-0.016
Williams ND	0.37	-0.57	2.51	-0.45	-0.096	-0.131	0.238	-0.03	-0.020
Kidder ND	-0.79	-1.12	-0.70	1.35	0.202	-0.256	-0.066	0.09	-0.025
Eddy ND	-0.58	-0.61	-0.20	-0.30	0.150	-0.139	-0.019	-0.02	-0.029
Perkins SD	-0.45	-0.41	0.03	-0.82	0.115	-0.093	0.003	-0.06	-0.032
Fallon MT	-0.14	-0.24	0.50	-1.05	0.037	-0.054	0.048	-0.07	-0.044
Billings ND	-0.35	-0.44	-0.86	0.70	0.089	-0.101	-0.082	0.05	-0.045
Towner ND	-0.57	-0.82	0.28	-0.52	0.146	-0.187	0.027	-0.04	-0.052
Fremont WY	1.09	0.76	0.73	-0.30	-0.280	0.174	0.069	-0.02	-0.058
Wibaux MT	-0.31	-0.56	-0.58	0.64	0.080	-0.129	-0.055	0.04	-0.059
Renville ND	-0.55	-0.69	0.04	-0.68	0.142	-0.158	0.004	-0.05	-0.060
Harding SD	0.14	0.35	-1.01	-0.18	-0.037	0.080	-0.095	-0.01	-0.065
Campbell SD	-0.35	-0.23	-0.80	-0.40	0.089	-0.052	-0.076	-0.03	-0.066
Richland MT	0.24	-0.05	0.69	-1.08	-0.063	-0.012	0.065	-0.08	-0.085
Oliver ND	-0.27	-0.30	-1.47	0.69	0.069	-0.069	-0.139	0.05	-0.092
Carter MT	-0.09	0.18	-1.28	-0.51	0.023	0.042	-0.122	-0.04	-0.093
McPherson SD	-0.61	-0.76	-0.43	-0.66	0.156	-0.173	-0.041	-0.05	-0.104
McHenry ND	-0.23	-0.16	-1.20	-0.30	0.059	-0.037	-0.114	-0.02	-0.114
Burke ND	-0.70	-1.07	0.37	-1.61	0.180	-0.244	0.035	-0.11	-0.143
Emmons ND	-0.52	-1.25	-0.07	-0.05	0.134	-0.285	-0.007	0.00	-0.162
McIntosh ND	-1.07	-1.96	0.74	-0.92	0.275	-0.449	0.070	-0.06	-0.168
Logan ND	-0.58	-1.34	-0.09	-0.29	0.148	-0.307	-0.009	-0.02	-0.187
Slope ND	-0.21	-0.33	-1.55	-0.46	0.055	-0.075	-0.147	-0.03	-0.199
Marshall SD	0.06	-0.85	0.83	-1.23	-0.015	-0.193	0.079	-0.09	-0.215
Dunn ND	0.34	-0.54	-0.63	-0.40	-0.088	-0.123	-0.059	-0.03	-0.299
Grant ND	-0.44	-1.56	-0.62	-0.02	0.112	-0.356	-0.059	0.00	-0.305
Sheridan ND	-0.66	-1.90	-0.29	-0.18	0.169	-0.435	-0.027	-0.01	-0.306
McKenzie ND	0.90	0.01	-0.58	-0.99	-0.232	0.003	-0.055	-0.07	-0.354
Roberts SD	1.13	-0.38	-0.61	0.81	-0.291	-0.087	-0.058	0.06	-0.379
Mountrail ND	1.09	-0.10	-0.03	-1.25	-0.279	-0.023	-0.003	-0.09	-0.393
Big Horn MT	3.09	-0.56	-0.03	1.68	-0.794	-0.127	-0.003	0.12	-0.806
Benson ND	2.55	-0.63	-1.15	-0.46	-0.655	-0.144	-0.108	-0.03	-0.940
Roosevelt MT	3.17	-1.03	0.99	-1.05	-0.815	-0.235	0.093	-0.07	-1.030
Corson SD	3.28	-0.65	-1.18	-0.51	-0.843	-0.148	-0.112	-0.04	-1.139
Rolette ND	3.92	-1.03	0.58	-0.59	-1.008	-0.235	0.055	-0.04	-1.229
Sioux ND	4.39	-0.61	-1.11	-0.43	-1.128	-0.140	-0.105	-0.03	-1.404

Appendix D

Leadership Survey/Interview Script

Introduction

Interviewer:
Thank you for participating in the interview. As you know, the interview will be tape recorded and a transcript will be created for use in my dissertation entitled, *Digital Deliverance: Dragging Rural Areas, Kicking and Screaming, into the Information Economy.*

The interview will consist of 15 questions. The first six questions will require a "yes" or "no" or designated response and they will be followed with a series of open-ended questions. At times during your response, I may ask you to elaborate more on a particular response, add an additional question, etc. Once the "YES/NO" or designated response questions are complete, seven open-ended questions will be asked. Please take as much time as you need to answer any of the questions. If at any time during the interview you do not feel comfortable answering a question or wish to stop the interview, please let me know and I will move on to another question or end the interview immediately. Do you have any questions about the interview process? Let's begin.

First Series of Questions

Interviewer:
Once again, please answer the following questions with a "yes/no" or another designated response and the question will then be followed by additional questions for you to elaborate upon your response.

Question One:
Are you directly involved in the __________________ information technology and telecommunications project/program?

YES	**NO**
If YES,	**If NO,**

What is your role? How did you hear about the project?
How did you become involved? What is your general knowledge of the
Are you compensated for your work?project?
How much time do you dedicate to the project weekly? How would you characterize your interest in the project?

Question Two:
Do you view the ________________________ project/program successful?

YES	NO
If YES,	**If NO,**
Why do you think it is successful? Any proof or measurement of success? How would you define success with regard to the project? What was the initial goal of the project from your perspective? In your opinion, are the goals of the project/program being met? Any examples?	Why do you think it is not successful? Any proof that the project is not successful? How would you define success with regard to the project? In your opinion, what or how could the project become successful?

Question Three:
Do you feel the community supports the __________________project/program?

YES **NO**

If YES, **If NO,**

(Same questions for either answer)
Why do you feel this way?
Any supporting evidence?
How would you characterize the community in general?
Do community wide project usually succeed or fail? Any examples or reasons?

Question Four:
Do you feel the _______________ project/program increased economic development activity in the community?

YES **NO**

If YES, **If NO,**

(Same questions for either answer)
Why do you feel this way?
Any proof or example from your perspective?

Question Five:
In your opinion, has the project been driven from the top down or the bottom up? Top Down meaning that a few community or project leaders have promoted the project to the community at large and Bottom up meaning the community at large accepted the project initially on its own merits (Interviewers note: Further clarification may be required)

TOP DOWN **BOTTOM UP**

If TOP DOWN, **If BOTTOM UP,**

(Same questions for either answer)
Why do you feel this way?
Any proof or example from your perspective?

Question Six:
Would you credit local leadership as the most critical factor in the success or failure of the ________________________ project/program?

YES **NO**

If YES,

Why do you feel this way?

If NO,

What would you say is the most important factor?
Why?

<u>**Second Series of Questions**</u>

Interviewer:
This series of questions are open for a free response. No "YES/NO" or designated response is needed. Occasionally during your response, I may ask a follow up question or ask you to elaborate on a particular point within your answer. Any questions? Let's begin.

Question Seven:

Has the program/technology assisted you in your profession? If so, how---if not, why?

Question Eight:

What type of information technologies, if any, do you use on a daily basis?

Question Nine:

Do you utilize high-speed telecommunications or direct cable for your Internet connection?

Question Ten:

Has the use of information technologies increased your profitability/productivity? If so, how?

Question Eleven:

Would you say that ____________________ community is one that is open to change? Why or why not?

Question Twelve:

Do you consider _________________________ community/city Urban or Rural? Why?

Question Thirteen:

Have or will you participate in any type of information technology related training? If so, where (did or will) you receive the training?

Question Fourteen:

If you could give one person a trophy for their involvement in the ________________ project/program, who would that person be?

Question Fifteen:

In your opinion, is there anything I forgot to ask about the project of importance? Any special comments on the project?

Conclusion

Interviewer:
This concludes the interview process. Thank you very much for your time and input into the research project.

References

Accordino, John J. 1994. Evaluating Economic Development Strategies. *Economic Development Quarterly* 8(1): 218–229.

Allen, John C. and Don A. Dillman. 1994. *Against All Odds: The Rural Community in the Information Age*. Boulder, CO: Westview Press.

Adkison, Jeff. 2001. A Wide-Angle View of the Location Process. *Area Development* 36(7): 48–51.

Applied Rural Telecommunications Clearinghouse. 1998. *The Role of Telecommunications in Rural Development*. Available on the World Wide Web:(http://www.aerie.org).

Arizona Department of Commerce. 2002. *ASPED to GSPED*. Position Paper. [cited 12 May 2002]. Available on the World Wide Web:(http://www.commerce.state.az.us/pdf/ miss/aspedGSPED.pdf).

Aristigueta, Maria P. and Jose I. Fernandez, Jr. 1998. Homegrown Economic Development: Implementing the Hispanic Business Initiative Fund. *Public Administration Quarterly* 22(3): 315–331.

Atkinson, Robert D. 2000. New Economy=New Criteria. *Business Xpansion Journal* December: 50–57.

Barrette, John. 2000. State Telecom Industry Argues Against Turf Encroachment by Public Power. *NebraskaStateNews.com*. [cited 12 December 2001]. Available on the World Wide Web:(http://nebraska.Statepaper.com/vnews/display.v/ART/2000/12/12/ 3a3577479?in_archive=1).

Bartik, Timothy. 1991. *Who Benefits from State and Local Economic Development Policies?* Kalamazoo, MI: Upjohn Institute.

Beck, Bill. 2001. Wiring Up for the Telecom Revolution. *Area Development* 36(4): 50–55.

Beck, Bill. 2000. Bridging the Digital Divide. *Area Development* 35(10): 22–29.

Berry, Brian J.L., Edgar C. Conkling, and D. Michael Ray. 1997. *The Global Economy in Transition*. New York: Prentice-Hall.

Berryhill, Ellery C. 2000. The Rural Reality. *U.S. Sites Relocation Journal* Spring: 18–23.

Bird, Dorreen Yellow. 2001. Leadership Development Expert Encourages Economic Growth in North Dakota. *Grand Forks Herald* 12 December 2001.

Blodgett, Jeffery W. 2000. *Information Technology: Creating Real Change for Small Business*. National Governors' Association Center for Best Practices Policy Paper. [cited 10 April 2002]. Available on the World Wide Web:(www.nga.org/Center).

Bonnett, Thomas W. 2000. *Competing in the New Economy: Governance Strategies for the Digital Age*. Philadelphia: Xlibris.

Bradshaw, Ted K., and Edward J. Blakely. 1999. What Are "Third-Wave" State Economic Development Efforts? From Incentives to Industrial Policy. *Economic Development Quarterly* 13(3): 229–244.

Bradshaw, Ted K., and Edward J. Blakely. 1979. *Rural Communities in Advanced Industrial Society: Development and Developers*. New York: Praeger.

Breathnach, Proinnsias. 2000. Globalisation, Information Technology and the Emergence of Niche Transitional Cities: The Growth of the Call Center Section in Dublin. *Geoforum* 31: 477–485.

Browne, William. 2001. *The Failure of National Rural Policy: Institutions and Interests*. Washington, DC: Georgetown Press.

Burress, David. 2001. The "Clusters" Challenge to Rural Development. *Kansas Business and Economic Review* 24(2): 1–5.

Cairncross, Frances. 2001. *Death of Distance: How the Communications Revolution is Changing Our Lives*. Boston: Harvard Business School Publishing.

Callanan, Brian. 2000. *Ireland's Shannon Study: Leaders, Visions, and Networks—A Case Study of Local and Regional Development*. Dublin: Irish Academic Press.

Caristi, Dom. 2000. Policy Initiatives and Rural Telecommunications in *Having All the Right Connections: Telecommunications and Rural Viability*. Westport, CT: Praeger Publishers.

Carney, Dan. 1996. Rural Legislators are Hoping to Plug into the Future. *Congressional Quarterly Weekly* 54(25): 1751–1753.

Castells, Manuel. 1985. *High Technology, Space and Society*. Beverly Hills: Sage Publications.

Castle, Emery N., ed. 1995. *The Changing American Countryside: Rural People and Places*. Lawrence: University of Kansas Press.

Castle, Emery. 1993. Rural Diversity: An American Asset. *Annals of the Academy of Political and Social Science* 529: 12–21.

City of LaGrange. 1999. *Intelligent City Application*. LaGrange, GA: City of LaGrange Economic Development Department.

Clark, David, Brian Ilbery, and Nigel Berkerley. 1995. Telematics and Rural Business: An Evaluation of Uses, Potentials, and Policy Implications. *Regional Studies* 29: 171–180.

Clarke, Susan E., and Gaile, Gary L. 1992. The Next Wave: Postfederal Local Economic Development Strategies. *Economic Development Quarterly* 6(2): 187–199.

Cleveland, Harlan. 1995. The Twilight of Hierarchy: Speculations on the Global Information Society. *Public Administration Review* 45.

Clemson University Community and Economic Development Program. 1996. Industrial Clusters Benefit Businesses, Spur Rural Manufacturing Job Growth. *The Community Leader's Letter* 7(3): 1.

Cohen, Stephen S., J. Bradford Delong, and John Zysman. 2000. Six Debates/Rules for a Digital Economy. *Blueprint: Online Journal of the Progressive Policy Institute* [cited 15 March 2002]. Available on the World Wide Web:(http://www.ndol.org/ndol_ci.cfm?contentid= 961&kaid= 07&subid=123).

ComMark. 2002. *ComMark: The Company*. [cited 22 April 2002]. Available on the World Wide Web:(http://www.commark.com).

Cooperative Association for Internet Data Analysis. 2002. *International Broadband Internet Backbone Map*. [cited 10 April 2002]. Available on the World Wide Web: (http://www.caida.org/tools/visualization/mapnet/Backbones/).

Coulson, L. Ann and David L. Darling. 1996. Coping Strategies of Successful Rural Communities. *Kansas State University Research Report 754* Manhattan, KS: Kansas State University.

Davis, Stan and Christopher Meyer. 1998. *Blur: The Speed of Change in the Connected Economy*. New York: Perseus Publishing.

Deavers, Kenneth L. 1989. *Rural Development in the 1990s: Data and Research Needs*. Research Paper from the Rural Social Science Symposium, AAEA, Baton Rouge, LA, July 28–29.

Dicken, Peter and Peter E. Lloyd. 1990. *Location in Space: Theoretical Perspectives in Economic Geography*. New York: Harper-Collins Publishing.

Drabenstott, Mark. 2001. New Policies for a New Rural America. *International Regional Science Review* 24(1): 3–15.

Drabenstott, Mark, and Katharine Scheaff. 2001. Exploring Policy Options for a New Rural America. *Main Street Economist.* Federal Reserve Bank of Kansas City. June.

Dslreports.com. 2002. *Broadband Frequently Asked Questions*. [cited 14 March 2002]. Available on the World Wide Web:(http://www.dslreports.com/faq).

The Economist. 2001. The Slow Progress of Fast Wires. *The Economist*: February 17.

Edwards, Michael. 1996. How Relevant is Development Studies? in *Beyond the Impasse: New Directions in Development Theory*. Schuurman, Frans. ed. London: Zed Books.

eircom. 1996. *Ennis Information Age Town: A Connected Community*. Ennis, Ireland: eircom.

European Union Committee of the American Chamber of Commerce. 2002. *The EU Committee Position Paper on the Promotion of Broadband in Europe.* [cited 11 March 2002]. Available on the World Wide Web:(http://www.eucommittee.be/Pages/fspop.htm).

Evans, Angie, and Mark M. Miller. 2000. Industrial Clusters and an Economic Development Tool. *Economic Development Review* 17(1): 48–56.

Falk, William W, and Thomas A. Lyson. 1991. Rural America and the Industrial Policy Debate, in *Rural Policies for the 1990s,* Flora and Christenson, eds. Boulder, CO: Westview Press.

Federal Communications Commission. 2008. *High Speed Services for Internet Access: Subscribership as of March 20, 2008.* [cited 11 August 2008]. Available on the World Wide Web: (http://hraunfoss.fcc.gov/edocs_public/attachmatch/DOC-280906A1.pdf).

Federal Communications Commission. 2002. *Inquiry Concerning the Deployment of Advanced Telecommunications Capability to All Americans in a Reasonable and Timely Fashion, and Possible Steps To Accelerate Such Deployment Pursuant to Section 706 of the Telecommunications Act of 1996.* Federal Communications Commission Report. CC Docket 98-146: February.

Federal Communications Commission. 2001. *High Speed Services for Internet Access: Subscribership as of December 31, 2000.* [cited 11 April 2002]. Available on the World Wide Web:(http://www.fcc.gov/Bureaus/Common_Carrier/Reports/FCC-State_Link/IAD /hspd0801.pdf).

Federal Communications Commission. 2000. *High Speed Services for Internet Access: Subscribership as of June 30, 2000.* [cited 11 January 2002]. Available on the World Wide Web:(http://fcc.gov/Bureaus/Common_Carrier/Reports/FCC-State_Link/ IAD/hspd1000.pdf).

Federal Reserve Bank of Atlanta. 2000. The New Economy at Work in Traditional Southern Industries. *EconSouth* 1: 9–13.

Fidelman, Miles R. 1994. *Life in the Fast Lane: A Municipal Roadmap for the Information Superhighway.* The Center for Civic Networking. [cited 10 December 2000]. Available on the World Wide Web: (http://civic.net/fastlane.html).

Fiorina, Carly. 2000. Building a New Corporate Structure. *Area Development* 35(3): 66–69.

Firestone, William A. 1987. Meaning in Method: The Rhetoric of Qualitative and Quantitative Research. *Educational Researcher* 16(7): 16–21.

Flora, Cornelia, and James A. Christenson. 1991. *Rural Policies for the 1990s.* Boulder, CO: Westview Press.

Florida, Richard. 2002. *The Rise of the Creative Class.* New York: Basic Books.

Florida, Richard. 2000. *Competing in the Age of Talent: Quality of Place and the New Economy.* Report for the R.K. Mellon Foundation, Heinz Endowments, and Sustainable Pittsburgh.

Foster, R. Scott. 1992. State Economic Policy: The Emerging Paradigm. *Economic Development Quarterly* 6(1): 3–13.

Freeman, Chris, and Francisco Louca. 2002. *As Time Goes By: From the Industrial Revolutions to the Information Revolution.* New York: Oxford University Press.

Friedmann, John. 1966. *Regional Development Policy: A Case Study of Venezuela.* Cambridge, MA: MIT Press.

Friedman, Thomas L. 2000. *The Lexus and the Olive Tree.* New York: Alfred A. Knopf.

Galston, William, and Karen J. Baehler. 1995. *Rural Development in the United States: Connecting Theory, Practice, and Possibilities.* Washington DC: Island Press.

Georgia Department of Industry Trade and Tourism. 2002. *Economic Development: What is it about Georgia?* Position Paper. [cited 12 May 2002]. Available on the World Wide Web at:(http://www.georgia.org/economic/index.asp).

Gibbs, Robert M., and G. Andrew Bernat, Jr. 1997. Rural Industry Clusters Raise Local Earnings. *Rural Development Perspectives* 12(3): 18–25.

Goff, Patrick. 2001. Closing the Digital Divide in McDermitt, Nevada. *Economic Development Review* 17(3): 100–101.

Goss, Earnest P., and Joseph M. Phillips. 1997. The Effect of State Economic Development Agency Spending on State Income and Employment Growth. *Economic Development Quarterly* 11(1): 88–96.

Grant, Elaine. 2002. Study: E-Commerce To Top $1 Trillion in 2002. *E-Commerce Times*: 13. [cited 3 April 2002]. Available on the World Wide Web:(http://www.ecommercetimes. com/perl/story/16314.html).

Grant, James. 1990. *A Long Term Assessment in Highlands and Islands: A Generation of Progress*. Aberdeen: Aberdeen University Press.

Hair, J.F., R.E. Anderson, R.L. Tatham, and W.C. Black. 1998. *Multivariate Data Analysis*. New York: Prentice-Hall.

Hansen, Devon, and Muhammad Hemmasi. 2001. The State of the Organization of the Islamic Conference at the Dawn of the New Millennium. *Prairie Perspectives Geographical Essays* 4: 258–282.

Hart, John Frazier. 1992. Nonfarm Farms. *Geographical Review* 82(2): 166–179.

Harvard Business School. 2001. *The Free Internet Initiative in LaGrange Georgia*. Harvard Business School Case Study: N9-302-41.

Haynes, Kingsley, Serdar Yilmaz, and Somik Lall. 1999. *Locational Effects of Telecommunications Infrastructure: Methodological Considerations*. Research paper from the American Association of Geographers Conference, Charlotte, North Carolina, October.

Hirschman, Alfred. 1958. *The Strategy of Economic Development*. New Haven, CT: Yale Press.

Hiss, Tony. 1990. *The Experience of Place: A New Way of Looking At and Dealing with Our Radically Changing Cities and Countryside*. New York: Vintage Books.

Hoepfl, Marie C. 1997. Choosing Qualitative Research: A Primer for Technology Education Researchers. *Journal of Technology Education* 9(1): 12–39.

Industry Ennis. 2001. EU Information Society Commissioner for Industry Ennis Seminar. *Industry Ennis News* (2): 1–10.

Innovisions Canada. 2001. *Definition of Teleworking*. [cited 12 May 2002]. Available on the World Wide Web:(http://www.ivc.ca/definition.htm).

James, Keecia. 2001. Arkansas: State of the State. *Southern Growth Policies Board Report*. [cited 11 January 2002]. Available on the World Wide Web:(http://www.southern.org /pubs/state/sf041201KJ.shtml).

Jarboe, Kenan Patrick. 2001. Knowledge Management as an Economic Development Strategy. *Reviews of Economic Development Literature and Practice*. Washington, DC: Economic Development Administration.

Johnson, Thomas G. 2001. The Rural Economy in a New Century. *International Regional Science Review* 24(1): 37. [cited 13 March 2002]. Available on the World Wide Web:(http://www.ebsco.com/online/direct.asp?ArticleID=CX0F9GK18AU1V5DEKBVV).

Kelly, Kevin. 1998. *New Rules for the New Economy*. New York: Penguin Books.

Kolzow, David, and Ed Pinero. 2001. The Internet Economy and Its Impact on Local Economic Development. *Economic Development Review* 17(3): 82–99.

Koo, Jun, Michael I. Luger, and Leslie Stewart. 1999. *Best Practices in Science and Technology-Based Economic Development Policy: U.S. and Global*. Chapel Hill, NC: University of North Carolina Office of Economic Development.

Korsching, Peter F., Patricia C. Hipple, and Eric A. Abbott, eds. 2000. *Having All the Right Connections: Telecommunications and Rural Viability*. Westport, CT: Praeger Publishers.

Kotkin, Joel. 2000. *The New Geography: How the Digital Revolution is Reshaping the American Landscape*. New York: Random House.

Kruger, Lennard G., and Angele A. Gilroy. 2001. *Broadband Internet Access: Background and Issues*. The National Council For Science and the Environment. [cited 18 May 2001]. Available on the World Wide Web:(http://www.cnie.org/nle/st-49.html).

Langer, Laura. 2001. The Consequences of State Economic Development Strategies on Income Distribution in the United States, 1976 to 1994. *American Politics Research* 29(4): 392–416.

Lillington, Karlin. 2002. Plan Will Bring Broadband Internet Technology to 123 Towns. *The Irish Times*: April 3.

Macklin, Ben. 2002. The Benefits of Broadband: The Costs. eMarketer.com. [cited 7 April 2002]. Available on the World Wide Web:(http://www.emarketer.com/analysis/ technologies/20020404_at.html).

Malecki, Edward J. 1997a. *Technology and Economic Development: the Dynamics of Local, Regional, and National Competitiveness*. Essex, UK: Longman.

Malecki, Edward J. 1997b. Telecommunications Technology and Business Location: A Review, in *Rural Infrastructure as a Cause and Consequence of Rural Economic Development and Quality of Life*. Schmidt, Joe, ed. Proceedings of a Regional Workshop, Birmingham, AL, 107–142.

Malizia, Emil E., and Edward J. Feser. 1999. *Understanding Local Economic Development*. New Brunswick, NJ: CUPR Press.

Maturi, Richard. 2001. Going Wired: The Rural Connection. *Area Development* 36(4): 78–87.

Mayer, Henry J., and Michael R. Greenberg. 2001. Coming Back from Economic Despair: Case Studies of Small-Medium Size American Cities. *Economic Development Quarterly* 15(3): 203–216.

McGrinn, Richard A. 2000, Communications Networks as Strategic Assets. *Area Development* 35(3): 70–74.

The McKensey Quarterly. 2001. *Pipe War: Which Broadband Technology Will Win the Race for Homes and Offices*. [cited March 2 2002]. Available on the World Wide Web:(http://www.mckenseyquarterly.com/article_page.asp?articlename=piwa01).

McKenzie County Job Development Authority. 2001. *Watford City: A Community with a Vision for the Future*. Watford City, ND: McKenzie County Job Development Authority.

McQuillian, Helen. 2001. *Ennis: The Information Age Town Evaluation Report*. [cited 30 May 2002]. Available on the World Wide Web:(http://www.ennis.ie/iat_evaluation_ report.htm).

Mitchelson, Ronald L. 1999. Spatial Aspects of Technology Transfer: Evidence from Kentucky. *Southeastern Geographer* 39(2): 206–219.

Moschella, David C. 1997. *Waves of Power: The Dynamics of Global Technology Leadership, 1964–2010*. New York: AMACOM.

Munnich, Lee, Greg Schrock, and Anna Hayes Bonelli. 2002. *Rural Knowledge Clusters: Implications for Minnesota State Colleges and Universities*. Research Policy Paper. Hubert Humphrey Institute of Public Affairs, The University of Minnesota.

Myrdal, Gunnar. 1957. *Economic Theory and Underdeveloped Regions*. New York: Harper and Row.

National Center for Small Communities. 1999. *Getting Online: A Guide to the Internet for Small Town Leaders*. Washington D.C.: National Center for Small Communities.

National Governors' Association. 2000. *State Strategies for the New Economy*. National Governors' Association Center for Best Practices.

National Telecommunications and Information Administration. 2000. *Falling Through the Net: Toward Digital Inclusion*. Washington, DC: U.S. Department of Commerce.

Nazam, Sufi. 1992. *Telecommunications and Economic Development in Rural America: A Case for Nebraska*. International Center for Telecommunications Management.

Neil, Cecily, and Markku Tykkylainen, eds. 1998. *Local Economic Development: A Geographical Comparison of Rural Community Restructuring*. Tokyo: United Nations University Press.

Nelton, Sharon. 1994. Bringing in the Jobs. *Nation's Business* (March).

Newsbytes.com. 2001. *Congressional Democrats Push For Rural Broadband Rollout*. [cited 29 May 2001]. Available on the World Wide Web:(http://www.newsbytes.com/news/01/ 166229.html).

Neiman, Max, and Kenneth Fernandez. 1999. Measuring Local Economic Development Policy and the Influence of Economic Conditions. *International Journal of Economic Development* 1(3): 311–339.

NUA Internet Surveys. 2002. Internet Access Growth by International Region. [cited 12 April 2002]. Available on the World Wide Web:(www.nua.ie).

Office of the Governor of Alabama. 2001. *State of the State Address*. [cited 12 February 2002]. Available on the World Wide Web:(http://www.governorpress.state.al.us/sp-2001-02-06-state-highlights.htm).

Odasz, Frank. 2001. *The Future of Community Development: Make the Living You Want Living Wherever You Want*. Lone Eagles Consulting Working Paper. [cited 11 May 2002]. Available on the World Wide Web:(http://lone-eagles.com/comdev.htm).

Office of the Governor of Georgia. 2001. *Governor Announces New Technology Initiative*. [cited 3 January 2002]. Available on the World Wide Web:(http://www.ganet.org/governor/press/press.cgi?prfile=PR.20010112.01&agencyid=AGENCYID).

Okagaki, Alan, Kris Palmer, and Neil S. Mayer. 1999. *Strengthening Rural Economies: Programs that Target Promising Sectors of a Local Economy*. Washington, DC: U.S. Department of Housing and Urban Development.

O'Hanlon, Dan, and D. Keith Comstock. 2001. *A Proposed Blueprint for the Use and Growth of Technology in West Virginia: Report to the Governor of West Virginia by the Technology in State Government Task Force*. [cited 3 January 2002]. Available on the World Wide Web:(http://www.state.wv.us/governor/tech.in.wv.htm).

Oxford English Dictionary. 1993. *The New Shorter Oxford English Dictionary on Historical Principals*. Oxford: Clarendon Press.

Parker, Edwin B. 2000. Closing the Digital Divide in Rural America. *Telecommunications Policy* 24:4. [cited 24 May 2001]. Available on the World Wide Web:(http://www.tpeditor.com/ contents/2000/parker.htm).

Parker, E.B., H.E. Hudson, D.A. Dillman, and A.D. Roscoe. 1989. *Rural America in the Information Age: Telecommunications Policy for Rural Development*. Lanham, MD: University Press of America and The Aspen Institute.

Perroux, Francois. 1950. Economic Space: Theory and Applications. *Quarterly Journal of Economics* 64: 89–104.

Peterson, Shane. 2000. Equal Opportunity Access. *Government Technology*. [cited 3 November 2001]. Available on the World Wide Web:(http://www.govtech.net/ publications/gt/ 2000/nov/ncFeature/equalOpp.html).

Phillips, Rhonda. 2000. Crossing the Digital Divide, Caribbean Style: Investing in Telecommunications for Economic Development. *Economic Development Review* 16(4): 20–25.

Phillips, Rhonda. 1996. *A Test of Business Growth through Analysis of a Technology Incubator Program*. Dissertation, Georgia Institute of Technology.

Plugge, Patricia L. 1996. Self-Help Strategic Planning for Small Communities, in *Practicing Economic Development*, Koepke, Robert ed. Chicago: American Economic Development Council.

Porter, Michael. 2000. Location, Competition, and Economic Development: Local Clusters in a Global Economy. *Economic Development Quarterly* 14(1): 15–34.

Porter, Michael. 1998. Clusters and the New Economics of Competition. *Harvard Business Review* November-December.

Porter, Michael. 1990. *The Competitive Advantage of Nations*. New York: Free Press.

Pulver, Glen C. 1998. *Issues in Rural Policy Development*. Fellows Seminar Paper. University of Wisconsin Center for Community Economic Development [cited 2 March 2002]. Available on the World Wide Web:(http://www.uwex.edu/ces/cced/ plvrfelo.html).

Quay, Ray. 1995. Telecommuting: Possible Futures for Urban and Rural Communities. *Arizona State University Online Planners Forum*. [cited 8 May 2002]. Available on the World Wide Web:(http://www.asu.edu/caed/onlineplanner/webtopic/ elecvilg/telecom.html).

Raagmaa, Garri. 2001. Public Leaders in Regional Economic Development. *European Planning Studies* 9(8): 1039–1055.

Romer, Paul. 1990. Endogenous Technological Change. *Journal of Political Economy* 98: 71–102.

Rosenfeld, Stuart A. 2004. Crafting a New Rural Development Strategy. *Economic Development America*, Economic Development Administration, Summer, pp. 11–14.

Rosenfeld, Stuart A. 2001. *Backing into Clusters: Retrofitting Public Policies.* Research paper from the John F. Kennedy School of Government Symposium, Harvard University, March 29–30.

Rosenfeld, S.A., C.D. Liston, M.E. Kingslow, and E.R. Forman. 2000. *Clusters in Rural Areas: Auto Supply Chains in Tennessee and Houseboat Manufacturers in Kentucky.* Carrboro, NC: Regional Technology Strategies.

Rosenfeld, Stuart A. 1995. *Industrial Strength Strategies: Regional Business Clusters and Public Policy.* Washington, DC: Aspen Institute.

Rowley, Thomas D., David W. Sears, and Glenn L. Nelson. 1996. *Rural Development Research: A Foundation for Policy, Vol. 170.* New York: Greenwood Publishing.

Ryan, George H. 1990. *Rural Revitalization: The Comprehensive State Policy for the Future.* Report of the Illinois Governor's Rural Affairs Council, April.

Schantz-Feld, Mali R. 2000. Texas Economic Development Profile. *Area Development* 35(11): 34–38.

Shannon Development. 2000. *Shannon Development Corporate Plan 2000–2002.* Shannon: Shannon Free Airport Development Company.

Smith, W. Thomas, and Steve Milano. 2000. High-Tech Industries Turning to Unlikely Locales. *Business Xpansion Journal*: October.

State of Florida. 2001. Governor Bush Launches PowerUP Florida: A Statewide Initiative to Bridge the "Digital Divide." *Office of the Governor of Florida.* [cited 1 December 2001]. Available on the World Wide Web:(http:/www.myflorida .com/myflorida/government/ learn/powerup/partnership.html).

State of Hawaii. 2001. House Bill 1617. *Hawaii State Legislature.* [cited 23 August 2001]. Available on the World Web:(http://www.capitol.hawaii.gov/site1/docs/ docs. asp?press1=docs).

State of Idaho. 2000. *Findings and Recommendations of the Governor's Task Force on Rural Economic Development.* September. [cited 23 August 2001]. Available on the World Wide Web:(http://www.idoc.state.id.us/news/ruraltask.html).

State of North Carolina. 2001. Senate Bill 641. *North Carolina General Assembly.* [cited 25 September 2001]. Available on the World Wide Web:(http://www.ncga .state. nc.us/homePage.pl).

State of Virginia. 1999. Chapter 916. *Virginia General Assembly Record.* [cited 25 September 2001]. Available on the World Wide Web:(http://legis.state.va.us/).

Stough, R.R., K.E. Haynes, R. Kulkarni, R. Auger, and M. Quingshu. 2000. *Industrial Clusters and Economic Development Issues.* Paper Presentation at the 23rd Applied Geography Conference, October 11–14.

Strauss, Anselm, and Juliet Corbin. 1990. *Basics of Qualitative Research: Grounded Theory Procedures and Techniques.* Newbury Park, CA: Sage Publications.

Summers, Gene F., Frances Horton, and Christina Gringeri. 1995. Understanding Trends in Rural Labor Markets in *The Changing American Countryside: Rural People and Places.* Castle, Emery N., ed. Lawrence: University of Kansas Press.

Sung, Tae Kyung, and David C. Gibson. 1998. *Knowledge and Technology Transfer: Levels and Key Factors.* Unpublished Paper. Austin, TX: University of Texas.

Swager, Ronald J., ed. 1991. *Economic Development Tomorrow: A Report from the Profession.* Chicago: American Economic Development Council.

Tassey, Gregory. 2002. *R&D and Long-Term Competitiveness: Manufacturing's Central Role in a Knowledge-Based Economy*. Washington D.C.: National Institute of Standards and Technology.

Thornburgh, Dick. 1998. A Path to Smarter Federal Leadership in Economic Development: Learning, Leveraging, and Linking. *Economic Development Quarterly* 12(4): 291–299.

Townsend, Anthony. 2002. *Urban Technology and Telecommunications Listserve*. [cited 12 March 2002]. Available in the World Wide Web:(www.informationcity.org/telecom-cities).

Traill County Technology Center. 2002. *Traill County Technology Center: Committed to Business*. [cited 12 March 2002]. Available in the World Wide Web:(http://www.trailtechnology.com/).

Turner, Colin. 2002. *The Information Economy*. New York: Kogan Page Publishing.

United States Bureau of the Census. 2000. *Population Statistics*. [cited 22 May 2002]. Available on the World Wide Web:(www.census.gov).

United States Bureau of the Census. 1995. *Definition of Rural*. Available on the World Wide Web:(http:// www.census.gov).

United States Congress. 2001. *Congressional Record Online*. [cited 12 July 2001}. Available on the World Wide Web:(http://www.access.gpo.gov/su_docs/aces/aces150.html).

United States Department of Agriculture. 2000. *Advanced Telecommunications in Rural America: The Challenge of Bringing Broadband Service to All Americans*. Research Report: Washington, D.C. April.

United States Department of Commerce. 1999. *Closing the Digital Divide*. Conference Speech. Washington DC: December.

United States Department of Commerce. 1998. *Falling Through the Net II: New Data on the Digital Divide*. [cited 2 March 2002]. Available from World Wide Web:(http://www.ntia.doc.gov/ntiahome/net2/falling.html).

United States Economic Development Administration. 1997. *Cluster-Based Economic Development: A Key to Regional Competitiveness*. Washington, DC: U.S. Department of Commerce.

United States House of Representatives. 2007. *Broadband Critical to Rural Development*. [cited 12 May 2007]. Available on the World Wide Web:http://www.house.gov/smbiz/PressReleases/2007/pr-05-09-07-rural-broadband.htm

United States House of Representatives. 2001. *Economic Development in Rural America, Small Business Access to Broadband Hearing*. [cited 12 May 2002]. Available on the World Wide Web:(http://frwebgate.access.gpo.gov/cgi-in/getdoc.cgi?dbname=107_house_hearings&docid= f:73403.wais).

Valley Metro. 2002. *Local Telework Case Studies*. [cited 13 May 2002]. Available on the World Wide Web:(http://www.valleymetro.maricopa.gov/Rideshare/Telework/local_ studies.htm).

Weber, Alfred. 1909. *Theory of the Location of Industries*. Cambridge, MA: MIT Press.

Webopedia. 2001. *Definition of Broadband*. Available on the World Wide Web:(http://www.webopedia.com).

Weng, Que. 1997. *Integration of Spatial Analysis and GIS for Spatial Economic Pattern Analysis*. Discussion Paper Series, Department of Geography, The University of Georgia.

West, Monieca. 1999. *Measuring Community Readiness for an Information Based Economy*. Thesis. Economic Development Institute.

Western Governor's Assocation. 1999. *CERA Strategic Plan*. Denver, CO: Western Governor's Association.

The White House. 2000. *From Digital Divide to Digital Opportunity*. Study from the Clinton-Gore Administration [cited 5 May 2001]. Available on the World Wide Web:(http://clinton4.nara.gov/WH/New/digitaldivide/digital1.html).

Williams, Malcolm. 1990. Leadership: The Key to Economic Development. *Economic Development Review* 8(3): 4–7.

Willoughby, Kelvin W. 2000. Building Internationally Competitive Technology Regions: The Industrial-Location-Factor Approach and the Local-Technological-Milieux Approach. *Journal of International and Area Studies* 7(2): 1–36.

Wohlbruck, Aliceann, and Melissa Levy. 2001. Linking Communities to Opportunities through Telecommunications. *Economic Development Review* 17(3): 34–39.

World Teleport Association. 2002. *What is an Intelligent Community?* Intelligent Community Forum Position Paper. [cited 12 April 2002]. Available on the World Wide Web:(http://www.intelligentcommunity.org/html/whatis.html).

Yankee Group. 2002. *The Yankee Group Predicts Continuing Growth in Broadband Adoption in North America*. [cited 12 May 2002]. Available on the World Wide Web:(http://www.yankeegroup.com/public/news_releases/news_release_detail.jsp?ID= PressReleases/news_may072002_globalnbna.htm).

Yin, Robert K. 1994. *Case Study Research: Design and Methods*. Thousand Oaks, CA: Sage Publications.

Young, Ruth C., Joe D. Francis, and Christopher H. Young. 1993. Innovation in Small Manufacturing Firms. *Community Development Reports* 1(4).

Youtie, Jan. 2000. Field of Dreams Revisited: Economic Development and Telecommunications in LaGrange, Georgia. *Economic Development Quarterly* 14(2): 146–153.

Index

www.ingramcontent.com/pod-product-compliance
Ingram Content Group UK Ltd.
Pitfield, Milton Keynes, MK11 3LW, UK
UKHW021034150926
475561UK00003B/105

9 780761 845522